T0155195

THE COMMUTER PIG KEEPER

A Comprehensive Guide to Keeping Pigs when
Time Is Your Most Precious Commodity

THE COMMUTER PIG KEEPER

A Comprehensive Guide to Keeping Pigs when
Time Is Your Most Precious Commodity

MICHAELA GILES PH.D.

5m Publishing

First published 2016

Copyright © Michaela Giles 2016

Published by
5M Publishing Ltd,
Benchmark House,
8 Smithy Wood Drive,
Sheffield, S35 1QN, UK
Tel: +44 (0) 1234 81 81 80
www.5mpublishing.com

A Catalogue record for this book is available from the British Library

ISBN 978-1-910455-53-1

Book layout by
Keystroke, Neville Lodge, Codsall, Wolverhampton

Printed by Replika Press Pvt Ltd, India
Photos by Michaela Giles unless otherwise credited

DEDICATION

I would like to dedicate this book to our two daughters Marcie and Connie. Both are no longer with us and the massive gap they left in our lives is the sole reason we strayed into the pedigree pig world. Marcie didn't get to meet any of our pigs as she died shortly after her birth, but Connie found our fattening weaners hysterically funny, so it seems fitting that we went further down the pig route. It doesn't replace them in any way, nothing ever could. It has, however, given us the opportunity to meet some fantastic people who have enriched our lives in other ways – people we will never forget, no matter where life takes us in the future.

CONTENTS

LIST OF ILLUSTRATIONS

LIST OF TABLES

FOREWORD

by Liz Shankland

The trend towards leaving the city behind and moving to the countryside is showing no sign of waning. If anything, the number of people migrating from urban areas to rural parts is rising, pushing the price of smallholdings up as they do so.

Back in the 1970s, when the pioneering smallholder John Seymour wrote *The Complete Book of Self-Sufficiency,* the emphasis was firmly on escaping completely from the rat-race and living entirely off the land. Today, few of us can afford to do just that. Most of us want the best of both worlds: to be able to embrace 'the good life' and enjoy living in a rural idyll, whilst still maintaining a decent standard of living. That, of course, comes at a cost.

Many of the people who attend the smallholding and animal husbandry courses I teach at Kate Humble's rural skills school, Humble by Nature, want to keep a foot firmly in both camps and plan to carry on with their day jobs – at least for a while – whilst juggling the responsibilities of running a smallholding. My job is to give them ideas of what might be achievable, to encourage them to think seriously before taking on too much too soon, and to help them make plans for the future.

Inevitably, by the end of each course, the majority of the group have decided that pigs will be their first four-legged purchase. Of course, pigs are so much more charismatic than any other farmyard species, and need no help in winning over aspiring smallholders. But they are such practical and versatile animals, too. You can enjoy the simple pleasure of rearing them, watching as your land gets rotivated and fertilised. Then, when they have completed their land management duties, you can fill your freezer with the most delicious and wholesome meat.

In this book, my good friend, Dr Michaela Giles – an experienced pig breeder who runs courses at her own smallholding and also has a very demanding full-time job – leads by example, outlining ways in which you, too, can keep the day job but also be a part-time pig farmer. Whether you go for the simple 'piglet to plate' approach and buy in pigs to fatten, or whether you plan to eventually start your own breeding herd, Michaela explains the importance of planning ahead and getting your pig paddocks set up correctly before you make your first purchase. She has a long, daily commute to work and knows, from experience, that fitting in pig-keeping duties around the 'proper' job requires good time-management skills, a lot of foresight to avoid unwanted disasters – and a robust sense of humour.

This is an unusual book, in that it is written by a seasoned pig keeper who also happens to be something of a boffin – so you know that all Michaela's recommendations regarding health and welfare are based on solid, scientific knowledge. At the same time, she understands perfectly how you – the Commuter Pig Keeper – will be thinking right now, as you embark on your first porcine adventure. Trust her knowledge, learn from her experiences, and, above all, enjoy your pigs.

Liz Shankland is the author of the Haynes Pig Manual, Haynes Smallholding Manual, and the Haynes Sheep Manual. She teaches courses for smallholders at Humble by Nature (www.humblebynature.com).

Liz Shankland with her Supreme Champion of Champions Tamworth in 2014.

Photo courtesy of Liz Shankland

PREFACE

If we had £10 for everyone who says, 'I have no idea how you do all this and work full-time,' then one of us (and it would be me) could probably work part-time. Then we tell them that we both have demanding careers and commute from near the south coast of England to central London/Surrey, and they start to think we are complete liars. Mention that we also run pig-management and consultancy services, and they take on a confused or accusing look, depending upon their personality. I then daren't let them know that I also regularly write for various magazines including the prestigious *Practical Pigs*; chair the West Sussex Smallholders' Club; am a school governor; act as the media secretary for the British Saddleback Breeders Club; hold the post of Regional Sector Representative for the British Pig Association; and as a family, exhibit and show our pigs nationally.

In addition to all that, I have written this book. I am not telling you this to show off – well, perhaps a little bit (humour me!) – but to try and demonstrate that if we can do it, then most other people with a passion and commitment could too.

It takes forethought, planning, sacrifice, commitment and a dollop of help. A few of your lunch breaks are taken up with piggy tasks, a few of your evenings may be spent catching up with work, but invest in a hands-free phone and you can get quite a lot done on the commute to and from work. The requirement to multitask increases exponentially with the number and type of pigs you own. The common problems are the same, but they just come around quicker and more frequently when you have more pigs.

Although many people dream of selling up and becoming completely self-sufficient, reality is often tougher. Living the 'good life' completely might not be physically or financially possible, but acquiring yourself a sizeable slice certainly is. It is more than possible to combine full-time work with keeping pigs, even with a commute; in fact it can be beneficial to the work–life balance. Want to know how? This book will tell you how we have organised our modestly sized pig herd, and how we manage to get everything done properly on the farm. There are, of course, other ways that would be equally acceptable and that would fit in with your lifestyle and family. Take what suits you from the book and adapt the bits that don't; as long as the pigs are happy and healthy, then all is well.

What I have tried to achieve here is a complete reference guide for small-scale farmers and hobby keepers to dip into as required, but with a greater depth in some areas than some other books offer. As a scientist by profession I have gone into greater scientific detail in some topics, in order to aid understanding, but, hopefully, I have translated the science into intelligent layman's terms. I feel this is important to enable the reader to understand how to adapt methods to suit themselves and their pigs. Although this book will be particularly useful to those who have to juggle their pig-keeping with full-time employment, the content is just as relevant for those fortunate enough to have retired.

ACKNOWLEDGEMENTS

There are numerous people without whose help I would have struggled to complete this book.

Firstly I would like to thank my husband Neil and my son Oliver who allowed me to bury myself away typing and who took on a greater share of the farm duties before and after work. How they coped without my supervision is puzzling.

Then in no particular order of the considerable quality and quantity of the help provided:

- Liz Shankland, for willingly endorsing my first book on pigs and for openly sharing her vast photographic collection. I consider this quite an honour from such a well-published author, journalist, writer, lecturer and TV personality.
- Rachel Graham of *Practical Pigs* and *Practical Poultry* magazines, published author and all-round English language boffin, for her help in 'subbing' my work – I never did understand entirely what that was but I am entirely grateful.
- Chris Graham, Rachel's assistant, writer, published author of many books and editor of *Practical Pigs* magazine, for allowing me to use many of the photographs from the magazine.
- Angela Johnson of Lucky George Farm and breeder of Large Black pigs, and Christopher Rowley, breeder of Goose Meadow pedigree KuneKune pigs – my invaluable USA 'go to' piggy people who reviewed all the USA-related parts for accuracy.
- Christopher Impey for knowing more about pigs than anyone else I know and freely sharing that knowledge with me, even if we occasionally disagree about some points.

Athena Clarke-Sheward for thinking of various alternative titles to the book, mostly when she was in recovery from her numerous episodes of jetlag due to developing an addiction for Los Angeles. My personal favourite was *Bringing Home the Bacon*.

I have met all of the people named above through keeping pigs and what a fantastic bunch of people they are.

Acknowledgements

INTRODUCTION

My name is Michaela and together with my husband Neil we run our modestly sized six-acre smallholding. We have a son, Oliver, who, when home from school, then college and now university, also helps out. Our farm has a nucleus breeding herd of pedigree pigs, numerous piglets, plus growing pigs of various ages. We also have a small flock of Welsh Mountain Badger Face sheep, an increasing-in-size herd of breeding Boer goats and a flock of Brahma chickens. Throw in a couple of horses, dogs and cats and you have our farm.

We specialise in breeding British Saddleback and Middle White rare breed pigs. Some we retain for local and national showing and breeding, others we sell as breeding stock; the remainder, bar a few to grow on for meat ourselves, we sell on as fattening weaners to other smallholders. From the farm we also host pig-keeping courses and pig-handling workshops, and we also offer a pig-pregnancy scanning service and consultancy. I won't lie; it has turned out to be one of the most pleasant ways for us to be able to afford such a fantastic herd of pigs, with every penny reinvested to allow for continual improvement.

Keeping pigs and working full-time can be easy for anyone in a nine-to-five job that doesn't involve a commute, and whose place of work is near enough to pop back at lunch-time. Equally, if you are working flexi-time or part-time, or working from home, you will find keeping pigs very easy. If you don't have to give much notice for annual leave then so much the better. Obviously if you are a long-haul pilot and you will be doing it on your own, then this book cannot help you at all. Throw in an interested partner or another pilot on opposite shifts, then keep on reading. As you will see from our careers, they are not known for their regular hours, especially Neil's; I am a post-doc scientist for the UK government, researching the mechanisms and cures of naturally occurring diseases in livestock – work which often requires travel abroad. Neil is a detective within a specialist squad of London's Metropolitan Police Service – not a job you can stop suddenly and pick up again in the morning. We both commute as well: door-to-door, my journey takes an hour each way on a good day and Neil's takes approximately two hours each way. I am fortunate in that I do get to work from home on different days each week. We run our farm together, and hopefully *The Commuter Pig Keeper* will explain how.

A SLICE OF THE GOOD LIFE

I'm not going to lecture on why I don't like eating pork from intensive farming production methods, as not everyone has an option to rear their own meat. Thankfully practices are changing to enable shoppers to vote with their purses and buy outdoor-born and -reared or organic pork. I think people are fed up with meat from dubious, untraceable sources, that may or may not be from multiple low-welfare countries of origin, without continually scrutinising the packaging.

There are a few commercial practices that I strongly object to, such as the use of farrowing crates for up to four months per year in the UK/EU and keeping sows in stalls for their entire lives in other countries. Watching a sow prepare for birth and then bond with her piglets is quite something and the thought of her being denied the chance to express such a natural maternal behaviour leaves me cold. Likewise, the teeth clipping and tail docking of piglets. I have never needed to perform either technique – proving it's a husbandry issue and not a pig issue. I have also never had the need to castrate and have regularly eaten entire-boar fattening pigs up to ten months of age without experiencing meat taint – although that might be a pig issue in that our slow-maturing traditional breeds might not be as susceptible. A lack of quality enrichment is another issue. Pigs are intelligent animals, so providing minimal enrichment must be like giving a professor the same children's book to read over and over again as 'entertainment'. No wonder they get bored and chew on each other's extremities. The UK has led the way within the commercial sector, with strict rules on the health and mental well-being of the pigs; it is getting better all the time in terms of teeth clipping, and tail docking is starting to decrease as the requirement to do so is reduced by more enriched environments; farmers now have to get permission from a veterinary surgeon to perform these procedures in the EU. Surgical castration without pain relief under seven days old is still sadly legal in the UK/EU, and although vaccinations to delay puberty are emerging onto the market, castration is still being performed. I also don't like weaning at a young age just to get the sow back into production; in the UK, weaning may occur as early as three weeks, but four weeks is the norm. However, shockingly, in the USA, weaning at one to two weeks old is permitted, although thankfully not the norm. There is an additional issue of the use of growth promotors; while the use of antibiotics as a growth promotor is banned worldwide, there are hormone-type products that are legally approved in some countries. The use of ractopamine to encourage fast growth of lean muscle in grower and fattener pigs is permissible in the USA, although, as customers wake up to the practice, and labelling of meat can now legally state it is not used, it should be hoped that the practice falls into decline.

The Soil Association UK and United States Department of Agriculture (USDA) certified organic pig production prohibits the use of nose ringing (to prevent rooting behaviour), restrictive farrowing crates, castration, tail docking, teeth cutting, growth promotors and also only allows pigs to be kept indoors under exceptional welfare conditions. In the USA, keeping indoors and castration are still allowed, but teeth clipping/tail docking, while not prohibited, is discouraged. Even rare breed traditional and heritage pigs are being successfully used in commercial and organic systems to improve quality – and this can only be a good thing.

If you are reading this book, then it is highly likely that you want to, or already, rear pigs yourself, because you want to know exactly how they have been treated, what they have eaten and what sort of life they have had. Add in the ability to select a specific breed with a superior taste and texture for less than it would cost to buy intensively reared pork at a supermarket, and it makes perfect sense.

Keeping pigs between the spring and autumn is how most people start; with the longer, lighter evenings and sunny early mornings it can seem quite romantic. Popping down to the animals after your 6 am coffee, feeding and topping up waters and checking all is well is positively pleasant. You get that 'no one else is up' feeling and it is just you and the pigs. The sheer quiet before you have to get the kids up, leap into the car or catch your train to work – there are few other such simple pleasures.

Not us though! We bought our first three pigs on a complete whim, just before winter set in. We went to collect them in our horse trailer (it was allowed in the UK then, but now species-specific trailers are required – see Chapter Two), and took them home – not having sorted out anywhere sensible for them to sleep! We tried turning our large compost bin onto its side and packing it with straw as an overnight bed, but every time one got in, the other two pushed it down the pen. So they had to spend their first night in a stable and Neil had to spend the next two days fencing a small pen for them and making a pig house. They were not very friendly to start with, so I took the time in the stable to start making friends with them, and within 24 hours they were happy to stand and be scratched. I was less happy when I noticed they had pig lice, but a phone call to the head of parasitology at work confirmed what the parasites were. However, he then asked me to *not* treat them, but pick them off and put them in a pot as he wanted them for teaching purposes! Picking off 100+ lice from semi-tame pigs without killing the lice was not quite so romantic, but on the plus side, by the time they were lice free they were very tame, even if I did feel ever-so-slightly itchy.

After a couple of days they went into their new outdoor home and we were very pleased that they were outside, as were they. It took approximately four days for the pen to turn from grass and wild flowers to mud. Our farm is on heavy clay (*never* keep pigs on heavy clay), so after each rainfall it turns into a cross between Superglue and Velcro. However, the pigs had a warm dry house and areas in the pen that weren't quite so like the Somme, and all was well. Another couple of weeks passed and we had a phone call from our neighbour that Starsky, Hutch and Huggy (don't judge us, you'll call your first pigs something daft as well) were chasing his horses and could we 'catch them' please. Fortuitously, it was the weekend, and a bucket of food soon got them back in, only for us to watch them get out again once the food was eaten. This time they made their way to the chicken house, scared the chickens out, ate the four freshly laid eggs in the nesting box and promptly fell asleep. So we discovered that an electric fence was necessary, and not just something that we had read was good at keeping them contained. We also read that barbed wire at snout level was good, but I didn't like that idea at all and we decided not to go down that path. So, another few pounds sterling poorer and we had a new energiser, battery and an electric-fenced pen for them. A few squeals on and they seemed happy to avoid the 'biting' fence and stay where we wanted them. We were also growing vegetables at the time and I can still clearly remember, after they had come back from the butcher's, eating a meal that consisted of our meat and our home-grown purple broccoli, courgettes and potatoes. How we managed to eat with such wide, smug grins on our faces remains a mystery.

Since those early days we have now worked out a system that suits us and the pigs, and we have found them the easiest livestock animals on the farm to look after. Of course you have to invest in a little time to maximise your experience, but the pigs have rewarded us many times over. Some people start off raising weaners once a year, and this is a fine way of trying out the different breeds and flavours that are available before you are experienced enough to commit to breeding, or finally choose what you think is the best meat or easiest breed to handle.

People who raise pigs from spring to the end of summer are perhaps the clever ones as winter can be a major drain on your sense of humour when you have quite a few pigs, and pen after pen is thick mud and watery puddles and the taps are all frozen for the seventh day in a row. As we keep pigs all year round, winter is something we have to accept, but it does test your commitment, sense of humour and even character on occasion. You have to be at a work meeting by 9 am sharp, it is at

TIP

Buy some decent over-trousers, a waterproof coat, gloves and hat – it seems a little less horrid if, when you take it all off, you are still dry and clean underneath.

least an hour's drive in rush hour, there is driving rain and it's dark – but the pigs need feeding. If you have a couple of fatteners in for pork, it is relatively quick, but you will need a shower before you get changed as you are also now muddy.

HOW TO ORGANISE THE WORK–PIG BALANCE

First of all, avoid as many rookie mistakes as you can – they all cost either money or inconvenience, but usually both – by finding yourself an experienced mentor or attending a decent pig-keeping course. You will easily save more money than the course costs by getting it right the first time, and do ask for a receipt as you could later claim it back as a business expense, should you become seriously addicted. There are a number of privately hosted courses running around the UK, so have a look at them all, not just the nearest, check their qualifications, affiliations and experience, and even call them up to see which approach would suit you the most. They should all offer some form of after-support, which I think is probably the most valuable asset you will take away. Some provide you with a manual or book, others only notes; some offer real hands-on experience such as picking up piglets, using a weigh tape, and an introduction to moving pigs with a board and stick, whereas others just let you go into the pens and stroke the pigs. Hopefully some form of accreditation, regulation or monitoring of the tutoring standards on private courses will be introduced soon, giving new pig keepers a guarantee of the accuracy of the information they receive. It is certainly a hoop that I would be happy to jump through if it allowed us to continue with our courses.

There are pig-keeping courses held at various agricultural colleges, but I'm not sure that anyone keeping pigs as a part of their job can appreciate the sheer dedication involved when they get to go home at 5 pm sharp to a free evening. You, on the other hand, will be lucky to be home by 7 pm if you commute, and then you have to head back out and do the pigs. Colleges also have teams of students to provide labour, so how can they possibly know what it's like to perform each task on your own, in the thick of winter, before and after work, every day? There are also free or low-cost pig-keeping seminars held intermittently around the UK by various organisations and agencies trying to bridge the gap between the commercial world and the small-scale/hobby farmer but some are better than others in terms of delivery, and you have to bear in mind that the people hosting these usually come from a commercially based knowledge background, so you may have to extrapolate what will work on a smallholding, which isn't always easy when you are a newcomer. Neither of the latter options provides an after-support service, which you need to take into consideration when deciding.

There are 'back yard keeper' pig-training courses in the USA provided by some individual states, County Agricultural Extension Offices, youth groups and/or linked to specific breed clubs and registries rather than privately advertised, as in the UK. Apart from pedigree breed standards and possibly identification rules for that specific breed, then the regulations, health care and husbandry skills will all be the same and worthy of attending.

LOGISTICS

If you have a house with land attached, or a large garden, then so much the better as it really does help if you can see to the animals in the morning without having to try and keep clean. If the land is 10 miles from your house in the opposite direction to work, then it is going to be difficult – not impossible, but difficult without decent help. So if you have yet to buy or rent any land, set yourself a maximum radius to search within, and try not to view land that would of course be perfect if it wasn't so far away. If you don't even look at it, you won't be tempted and succumb to taking on something that you know will be impractical.

NOTE: When viewing land to rent, make sure before you sign that the landowners, and any local commercial pig farms, are happy for you to keep pigs, as some are not. If you are buying, check that there are no covenants preventing pigs on the land. In the USA check the property and land you are viewing is within an agricultural zone and keeping pigs is permitted.

So once you have your land, how do you set it up? The land should be organised with time management in mind. This can be almost irrelevant if you only have one pen with a couple of weaners or growers in it, but if you have quite a few it becomes very important. The last thing you want is lots of pigs acres away from each other with no water supply close by. To be a successful *Commuter Pig Keeper*, time is your most precious commodity and cannot be squandered.

At the time of writing we have 16 adult pigs, ten potential breeding/show gilts and a few fattening pigs, so at certain times of the year we can have up to 100 pigs at any one time if you include the nursing piglets. This is probably the maximum we would want to keep before it risked becoming a chore that we didn't enjoy. We have our land divided into four fields, with one of them next to woodland. In addition we have two small farrowing pens and one isolation pen near to the house. The field next to the woodland is now the 'pig field' and all our pens have been set up around the edge of this (see Figure 1.1).

On one side is a water tap with the longest hose in the world attached. In the chapter on husbandry, you will see why, although tempting, we don't use automatic drinkers. We start our feeding at the pens nearest to the tap in either a clockwise or counter-clockwise direction and then come back to the tap. We then pick up the hose and repeat the process but topping up the water troughs. In icy conditions, we also drain the hose after each use and hang it off the ground so that it is ready to go as soon as the tap is working, and fill up water butts outside each pen in the evening, once we

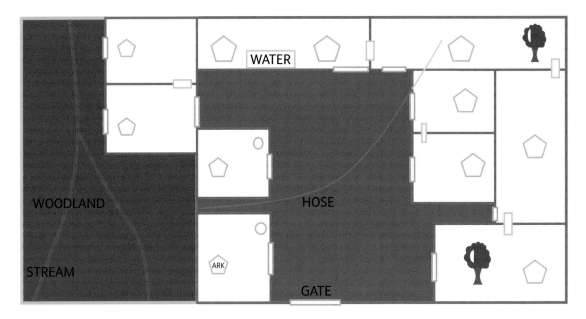

FIGURE 1.1 The set-up in the one-acre paddock with access to another acre of woodland.

get the tap working, so that if the tap is frozen in the morning we can top up the troughs using a bucket. This removes any worry about the pigs being left with no water all day and reduces the amount of flapping in the morning when our time pressure is at its greatest. Piglets need to have access to water from two weeks of age, by law. Deep water troughs, which are perfect for adults, as they are too heavy to tip up, are not an option for piglets, as even if they could get in they would not be able to get out and would very likely drown. We use a range of shallow troughs designed primarily for sheep, and buckets that fit into tyres for slightly older piglets, with the tyres semi-buried so they cannot easily tip them over. In the height of the summer we provide at least three troughs just in case they overturn them, fill them with mud or decide to sleep in them, ensuring the provision of water all day. We also have a kind neighbour who will check all the water levels during the day during very hot spells and top up if required. It is quite reassuring when she confirms that they are all fine and didn't need any extra water. The water troughs are all sited near the edge of each pen within reach of the hose and, where possible, one large trough covers two pens. The idea is that you don't have to climb into each pen to top up the waters.

The siting of their housing is also important. At least once a day in the winter you need to check that the bedding is still dry; which means entering each pen and tramping through thick mud, so the easier it is to reach the better, but this is covered in more depth in the husbandry chapter. We usually do it in the evening while each trough is filling with water – a watched trough seems to fill very slowly, so you might as well do something useful.

A Slice of the Good Life

From spring through to autumn/early winter we put as many sows and gilts into the woodland as possible – tips on reducing squabbles when you introduce them is covered later. We would leave them in there all year but we have a stream running through it which is prone to flooding and we don't want to have to dig out pregnant pigs from a clay-bed stream. In the summer we don't have to worry about shade from the sun as there is plenty available for all the woodland-kept pigs. Pigs can get sunstroke, and sunburn if they have any pink skin, so shade in the summer is absolutely essential and one of the 'Five Freedoms' of animal welfare that you are legally obliged to provide. It also means we can rest the heavily used winter pens.

For pigs about to give birth we have a choice of farrowing pens up by the house and pens next to the woodland, where the overhanging trees provide shade. As we have to keep boars as far away from each other as possible, at least one boar pen without natural shade is inevitable for us, so we have a range of artificial shading in place. Old patio umbrellas, four posts in the ground with a top and tarpaulins have all been used with varying degrees of success and aesthetic appearance. Remember that if the shade is broken or taken down by the pigs when you visit in the morning, and it's going to be a hot day, it can add 15 minutes to the time, as you will have to fix it. Try to use natural shade where possible, e.g. siting your pen around a tree, or make your shade as robust as 'piggily' possible. Another great sunscreen tip is to provide a wallow, which is a large muddy puddle; actually if you don't provide a wallow in a place of your choice then the pigs will make their own in a place of their choice, using water from their trough. This will turn the area into a complete bog and will deplete their precious drinking water. If they have access to a wallow, they are less likely to lose water through their skin, so they will drink less, which in turn means their water is more likely to last the entire working day.

Another time-saving tip is to fill your wheelbarrow in the evening with the sacks of feed you will need in the morning – assuming you don't have problems with vermin such as rats and squirrels. Knowing how much your scoop holds in weight of feed is useful, too, so you can just throw the appropriate number of scoops into each pen. Fill up every water trough in the evening to the brim, even if it's driving with rain and pitch black, and always leave the open end of your hose in the same place so you know exactly where to find it in the morning – but do not leave it in with the pigs or they will chew it to bits. Order more feed when you get down to a certain number of sacks, or if you buy on a smaller scale make sure you have a routine for when you visit the agricultural store; at the beginning of each month buy all the livestock feed you need for the month, as this way you only have to think of ordering and collecting feed 12 times a year and you are also less likely to overfeed. If you know your feed stocks are accurate to last for a set period, then the temptation to give an extra handful to a pig begging for more will be reduced – twice daily those extras can add up quickly. The pigs will remain at the correct weight and be healthier, giving you better-quality meat for less money.

If there is more than one of you running the operation, then communication and establishing standard procedures are essential. As we have a range of things to do on the farm – selling pigs, website updates, ordering pig food and straw, sorting out routine vaccination and deworming, the pig-keeping course bookings, providing after-support to our delegates, account paperwork, enquiries and emails to answer – it is essential that one person is in control of each aspect or mistakes are made and problems arise, quicker than you think. So we assign a lead person who has the responsibility for each task; if they are away, they hand over to the other person. This is easier when it's a couple living together, but what if your partner is a friend, who doesn't live on site? If you always do the morning feed and then, one day, you forget to tell them that you have an early start, your pigs will definitely not be fed that morning and hungry pigs are more likely to escape. They may also run out of water and not be checked for injuries, all of which risks souring the partnership and takes pleasure away from the pigs.

We also have set days for performing certain tasks. Although bedding is checked every day to make sure it is dry, Saturday is our preferred day to scrub and refill the water troughs, and while they are refilling we use the time to top up the bedding in the arks (pig housing). Some of the troughs may need another go during the week, but at least once a week they are cleaned out thoroughly. Start thinking about how you can design time-saving processes into your pigs' care because, as an added bonus at the weekend, on those heavenly days off and during the summer evenings, you can use all of the extra time you have made to interact with your pigs, learn their quirks and foibles and truly enjoy their company.

Another time-saving consideration, believe it or not, is your choice of breed. We used to have Gloucestershire Old Spots, which have great difficulty seeing where they are going, and we thought ours weren't the sharpest tools in the box either as an added difficulty. We could waste a good five to ten minutes trying to find them in the woods at feed time. They heard us coming with the wheelbarrow, they were as excited as the other pigs about their breakfast, and then they promptly marched off in two opposite directions. The others by then were all tucking in, so if they didn't get their feed they would miss out on the whole ration, and so you had to find them. It was very frustrating! The other breed we thankfully only ever had one of is the Mangalitza. She seemed to be able to tolerate a zap from the electric fence in the winter due to her woolly coat, which meant if our outer stock fencing wasn't as tight as a drum she could go from pen to pen causing havoc and eating twice as much as she should. Just a thought when choosing breeds – look up their general characters as well. Of course you may decide that the positive attributes of a breed outweigh the negative for you and your system.

You have to be realistic *before* you get pigs; if someone cannot see them twice a day, then don't get pigs; if money is already tight, then appreciate that pigs are expensive to feed and house and there is always the possibility of a vet bill, which again is far from cheap. I have seen pigs reared 'on the cheap' and it is unpleasant viewing. I have seen sows and piglets housed in cobbled-together pallets; pigs fed fruit and vegetables way past their best, even mouldy, and others fed countless loaves of bread, cake, stale milk and raw eggs as a regular part of their diet. Believe me, it affects the quality of the meat and even the animals' temperament when handling them. It's quite simple – if you cannot afford them, then don't buy them, and if you are not prepared to feed them properly, then don't buy them. Also, be realistic on the return you will get from your investment and don't think you can give up your day job or order an Aston Martin. The best return you will see at first is if you buy three piglets and sell two as butchered meat to the end user, and then this should pay for your meat from the third pig, and off-set some of your set-up costs. I will go into costs in more depth in Chapter Eleven.

I will now assume that you have enough money to feed and house your pigs. I will also assume that you and/or your team can visit the pigs twice a day. Depending upon how many pigs you have, then there will of course be some sacrifice. Watching TV can be curbed somewhat, with the amount you watch inversely proportional to the number of pigs you have. I will never win an award for the cleanest house – I probably wouldn't if I had no animals at all, but now I have an excuse. Holidays . . . I will have to Google that one as I have forgotten what they are, but in truth because we work long hours and often work abroad. I particularly like staying at home anyway. We do know other pig keepers who have regular holidays, using the bribe of pork to piggy sitters with some success, and there are professional farm sitters who advertise their services, but remember they are not cheap and could wipe out any profit or free meat from that third pig.

If you only have weaners once a year, then you can time your holiday to start after you have sent them off to the abattoir. As long as your butcher knows what you want, and you have worked out how to get them to him, then you can sun yourself knowing that your pork will be waiting for you when you return, or time it so that the meat is already sold on or in your freezer before you go.

Another time-saving tip is to involve any children you have access to; it's a great way to introduce your children to the realities of meat. You still need to oversee the proceedings dependent upon their level of maturity, but they can feed and top up waters while you do the observing. My advice is to be honest and explain that we give them the best life we can and then we eat them. It can backfire, as our own son would not eat any of the meat we produced for quite a few years. I think it was genuine to start with, when we only had a few each year and they all had names, but there is no way he could tell who was who now and yet he kept it up for appearances' sake in my opinion. He really missed out on the most legendary sausages. Most children accept it, and although they may protest at first, they soon forget if you don't make a huge deal out of it. Other halves can be a bit the same as well if they don't share the passion, but persevere – they won't recognise who is who when they are butchered.

If you are prepared to make a few sacrifices and can make the time, then go for it, but be warned it is an addictive game. We started with a few weaners per year and now have an all-singing, all-dancing herd of the most gorgeous pigs. In fact my husband will not be happy until my horses (the biggest parasites on the land, according to him) and sheep are all balancing on a narrow strip of grass! I turn my back for a minute and another ark has been delivered and a new pen built.

All you need now is to know how to do it correctly, which should be achieved within the *Commuter Pig Keeper* chapters. I will put in all the time-saving tips we employ as I go along, and include them in the relevant areas.

GET YOUR PAPERWORK IN ORDER

All countries will have some form of rules and regulations that will need to be adhered to. In the UK, the rules and regulations you have to abide by are the same for you as they are for the commercial guys, but they are sensible and not too onerous. So make sure you know what they are and comply, because once you start keeping pigs, you can be visited by numerous authorities with little notice to make sure all is well. Some want to see the health of the pigs, others want to check your paperwork, and some will do both. In the UK, Trading Standards are likely to be among your first visitors, but we have always found them helpful, and even useful, so don't panic.

REGULATIONS IN THE UK

Agricultural Land Identification

All land used for agricultural purposes – even a large garden – has to be registered with the Rural Payments Agency (England), Rural Inspectorate (Wales) or Rural Payments and Inspections Directorate (Scotland). The land will be issued with a County/Parish/Holding (CPH) number, which is made up of nine numbers, e.g. 12/345/6789, and uniquely identifies the land. The same number is used for all livestock kept on your land and is free of charge to obtain.

NOTE: When looking for properties to buy and you notice 'registered smallholding' in the brochure or advert, all that means is that it has a CPH number already. If it isn't already registered, since it's so easy and free to do, it's not a deal-breaker on the sale.

Becoming a Registered Keeper

Once the first set of pigs arrive on your holding, within 30 days you will need to apply to the Animal and Plant Health Agency (APHA) to register as a pig keeper and you will be issued with a *herd number*. A herd number consists of one or two letters and four numbers, e.g. AB1234 or A1234. This number is a unique identifier for the herd of pigs you own; even if you have gaps in time when you have no pigs, the number will be the same when you get pigs again and you don't need to reapply for it. It is this number you will use on your pigs' ear tags or slap-marks (shoulder tattoo) when they go to the abattoir, even if you purchased them from someone else. It is also used as a part of your secondary identification ear-tag number if you start pedigree breeding; if you move a pig over 12 months of age from your farm to another farm; to a show or exhibition at any age; and on your electronic Animal Movement Licences.

Movement Licences

There are rules about moving pigs from A to B that must be complied with by law. It is sensible, in that if there was an outbreak of a disease in an area, then all pigs that have moved out of that area or into it can be traced and monitored for signs of the disease in question. It can help stop the spread of disease quickly, prevent farmers from losing valuable bloodlines and save taxpayers millions of pounds. There is, of course, a downside: if the disease in question is a 'notifiable' disease, the government can order the humane destruction of your pigs. If the pigs are registered as pedigree, you will receive more compensation than if they were cross-breeds, since the impact on a pedigree breeding herd would be higher and it would cost a lot more to re-establish.

All the movements of livestock on and off your land are monitored by government agencies with the help of Animal Movement Licences (AML) – an online movement system. If you live in England or Wales, you will need to register as a producer for eAML2 (www.eaml2.org.uk, Figure 2.1), or if you live in Scotland you will have to register with ScotEID (www.scoteid.com).

Each movement of pigs requires prior notification. Although the online systems are in place, pig keepers can also give prior notice of a movement by telephone or post. No specialist software is required to use the online service. The seller sets up a movement, filling in details of the departure premises and the pigs' destination. The movement licence will also

FIGURE 2.1 The front page of the eAML2 electronic movement licence website.

include the number and type of pigs being transported, journey times, the name and vehicle registration of the haulier, biosecurity/health information, the country where the pigs were born and reared, and the system in which they have been reared (controlled or non-controlled – see below). If the journey is to an abattoir, there will be Food Chain Information (FCI) detailing any medicines administered within the preceding 28 days.

The controlled or non-controlled housing conditions, set by the EU and enforced by the Food Standards Agency, seem slightly ambiguous in their wording. Controlled housing is where the risk of coming into close contact with rodents, foxes and other wildlife capable of harbouring and transmitting the parasite *Trichinella* is considered low. Conversely, non-controlled is where that risk is considered high. Outdoor-reared animals are the least easy to define. The FSA says if you can perform a risk assessment showing minimal interaction between your pigs and *Trichinella*-susceptible wildlife, then you *can* call your husbandry system 'controlled'. The Agricultural and Horticultural Development Board (AHDB) Pork has produced some guidance on assessing the risk and specifying whether controlled or non-controlled may not be a requirement by the time you read this book, depending upon the results found between 2014 and 2017.

There are numerous 'drop-down' movement options to choose from, but the most commonly used ones would be from farm to farm, farm to abattoir or farm to show. Entering the details generates the appropriate haulier sheet to print off, which *must* accompany the pigs on their journey. It is likely that the haulier form will be available in smartphone format at some point. The seller indicates the pigs have left the farm and, when the pigs have been received at the destination, the buyer confirms the movement.

If you haven't registered for eAML2 or ScotEID before you collect your first pigs, the seller can set you up on the system, provided you have a CPH number. When the haulier form is printed out, it will come with a letter generated by the eAML2 or ScotEID systems, telling you what to do next. Registration must be completed within 30 days. Every incoming and outgoing movement of your pigs is held in a movement archive on the system.

NOTE: 'No CPH = No pigs.'

The first time you initiate an animal movement licence on the electronic system yourself will most likely be to the abattoir. You will also be asked a couple of questions that you will not know the answer to; remember the movement forms are for the commercial guys as well, who have to jump through additional legal hoops, especially if they are within an Assurance Scheme. If you are not sure how to answer a question, ask a breeder, pig course provider/mentor or give the eAML2 or ScotEID people a telephone call or email if not urgent.

There is also a statement that says, 'To comply with Welfare of Animals Transport Orders (WATO) transporters must carry an Animal Transport Certificate.' Don't panic – there are exemptions that I explain later.

Movement – Haulier

The haulier transporting the pigs may have to comply with the WATO rules if the movement is connected to an economic activity. The regulation does not define what constitutes an economic activity, but says: '**Transport for commercial purposes is not limited to transport where an immediate exchange of money, goods or services takes place. Transport for commercial purposes includes, in particular, transport which directly or indirectly involves or aims at a financial gain.**' So that's nice and clear, then! The full WATO conditions and exemptions have been condensed into a convenient document, and a website link is listed in Useful Contacts.

If it's *not* an economic related journey, then the eAML2/ScotEID-generated haulier sheet is all you need to move your pigs. If your journey *is* economic but less than 65 km (approx. 40 miles), then the haulier sheet is all you need to move your pigs, and you can skip the next bit.

If your journey *is* linked to an economic activity and *over 65 km* but less than eight hours in duration, then you will need a two-part authorisation licence, both parts of which must be carried together.

The first part is called a Welfare in Transport – Animal Transporter TYPE 1 Authorisation Licence and is available from the APHA. It is currently free of charge and remains valid for five years. The second part you will require is the LEVEL 2 Certificate of Competence in the Welfare of Animals in Transport (WAIT), which you can only get by sitting a paid-for examination. To obtain the LEVEL 2 Certificate of Competence you will need to pass a National Vocational Qualification (NVQ) examination. These exams are held all over the UK, usually in agricultural colleges, and can optionally include all species of large livestock plus horses. On passing you will receive a certificate, which you must carry on all journeys along with the TYPE 1 authorisation licence. Some colleges also host online WAIT courses and examinations.

NOTE: There is a Level 1 Certificate of Competence which covers you for journeys of over eight hours and also requires your livestock transportation vehicle to be inspected; it is probably outside a *Commuter Pig Keeper's* requirements and so is not covered here.

Trailer Regulations

Rules have also been implemented regarding livestock trailer suitability. The trailer must have a ramp no steeper than 20°, have loading gates fitted, have non-slip flooring, and it must be easily cleanable and disinfected. It must offer protection from the weather – so a solid roof is essential – and it must have appropriate ventilation. Pigs must be able to comfortably stand up and lie down in their own space and they must be provided with suitable bedding (e.g. straw) during travel, which is both absorbent and insulating.

There is talk of further criteria being introduced in future to ensure that all the pig waste created during a journey will to be contained in a sump tank and then disposed of back on the farm, but this is currently not a requirement. Before you buy a second-hand trailer, make sure it complies fully with the current laws.

General Good Practice

The WATO guidance regarding animal welfare is very clear, whatever type of journey you are undertaking. It says: 'No person shall transport animals or cause animals to be transported in a way that is likely to cause injury or undue suffering to them.'

Anyone who takes animals on a journey, whatever the length, should always follow good transport practice:

- the journey should be properly planned and time kept to a minimum;
- the animals must be checked and their needs met during the journey;
- they must be fit to travel;
- the vehicle and loading and unloading facilities should be designed, constructed and maintained to avoid injury and suffering;
- those handling animals must be trained or competent in the task and should not use violence or any methods likely to cause unnecessary fear, injury or suffering;
- water, feed and rest must be given to the animals as needed;
- sufficient floor space and height must be allowed.

Any derogation from these rules can only be given by a qualified veterinary surgeon. It is also prohibited, again except under qualified veterinary advice, to move pigs in their expected final 10% of gestation (11.5 days) and for seven days after birth; plus piglets aged less than three weeks must not travel more than 100 km (62 miles).

Standstill Rules

The minute the pigs arrive on your holding, those animals plus any existing pigs on your holding may not be removed, except to the abattoir or under the direction of a qualified veterinary surgeon, until 20 days have lapsed. This is known as a 20-day standstill.

NOTE: Pigs born on your holding do not initiate a standstill.

In the UK, if you have any existing cattle, goats, sheep or farmed deer on your land, they will be subject to a six-day standstill (13 days in Scotland). More livestock can come on to your holding, but a new standstill is triggered each time. If the new arrival is a pig, existing pigs cannot move for another 20 days and existing sheep/cattle/goats/deer

TIP

Write the end of the
20 days down on your
calendar so you don't have
to keep looking it up. Not
so important with a couple
of weaners, but if you have
a herd of pigs, it becomes
very important.

cannot be moved for six days (13 days). If more sheep/cattle/goats/ deer move onto your holding, then all animals are subject to a six-day (13-day) standstill, including the pigs – unless the six days (13 days) falls within an existing 20-day standstill. Read it a few times: it does make sense, I promise!

Standstill Exemption

If this is very inconvenient, and you have a suitable area away from your other animals, then you may apply for a Department for the Environment, Food and Rural Affairs (Defra) approved isolation facility. This involves form-filling by a Defra-licensed veterinary surgeon and most livestock vets will bring the application form with them on the visit. The easiest way of finding out which vets are registered with Defra is to ask your local APHA field office or ask your own livestock veterinary surgeon. Pigs moved onto the farm can then be held in the facility for the full 20-day standstill (or other species for 6/13 days) and the other livestock on the farm are not affected. At the time of writing, this is free of charge and Defra pays the veterinary surgery directly.

NOTE: Defra and the Animal Health and Welfare Board for England are currently considering relaxing the standstill rules, so keep an eye on the Defra and APHA websites.

Movement – Identification of Pigs

Farm-to-farm Movement for Fattening Pigs

When you buy a pig less than 12 months old, it can be moved from the holding where it was born to your holding with a temporary mark, e.g. a coloured spot or stripe made by a spray marker. The only legal requirement is that it must last until the end of the journey. If you are keeping the pig, then it will not require an ear tag until it moves off your holding.

Farm-to-farm Movement for Pedigree Breeding Pigs

The law here is exactly the same as for fattening weaners. However, when buying pedigree registered pigs, they will need to have an ear tag bearing the herd mark of the original breeder, as it forms a part of their pedigree identification alongside any notch or tattoo requirements the breed may have. Registered pedigree pigs have a spe-cial derogation in law, meaning they can be moved using the original breeder's herd mark and identification numbers, but if their final movement is to an abattoir, they will additionally need to carry your herd mark (see below) and it is illegal to remove an ear tag once applied, except with the permission of a licenced government official.

Farm-to-abattoir Movement

Ideally, when you buy weaners specifically for fattening, you don't want them ear tagged by the breeder but moved on a temporary mark – as mentioned earlier. You have three identification options to choose from for movement to the abattoir – an ear tag, a slap-mark (a tattoo of your herd mark) on each shoulder, or an ear tattoo. The identification must be easily visible when the pig is alive; so if you choose to slap-mark for carcass identification on a coloured pig, then you'll also have to insert an ear tag. Ear tags must have your herd mark printed or etched on, prefixed by the letters UK, and be of such quality that they can still be read after processing – so they need to be made of metal or a heat-resistant plastic, but not hand-written in ink. An individual number isn't required, but some breeders like to have individual numbers for their own reference. Slap-marking does not require the prefix UK and by law must only be applied once to each shoulder. Ear tattoos (not to be confused with pedigree tattoos) can be applied to one ear and UK prefix is optional.

NOTE: When purchasing pedigree fattening pigs and you require traceability to produce pedigree meat certificates, the full pedigree identification rules still apply in addition to the legal criteria above.

Herd Register or 'On-Farm Movement Records'

So you have picked up your pigs or had them delivered to your holding. After you have acknowledged their arrival on the eAML2/ScotEID system, the pigs' movement details are stored in the movement archive. If you only ever buy in pigs and send them to the abattoir, the movement archive efficiently details all the pigs on your holding at any one time and how they entered and left your holding, so this can be used as your herd register.

If you start breeding, you will need an additional register of any piglets born on your farm. The *information* is the legal requirement, not how you store it, so if you are a computer whizz then this is OK, but I would keep an up-to-date hard copy as well, just in case. The records must consist of the pig's identification, sex, date of birth/age, holding of birth, the date it moved on and/or off your holding, the destination if moving off, and a rolling tally of pigs on site. A record must be entered within 36 hours of the event happening and must be kept for inspection for up to six years.

An annual inventory of pigs on your holding, on a specific date of your choice, must also be recorded, plus a total of the normal number of pigs kept. This must be kept for three years after you stop keeping pigs and made available to government inspectors on request.

Feeding Pigs and the UK Law

You need to be aware of the current legal requirements on feeding your pigs. Long gone are the days of feeding kitchen waste (swill) in the UK, and pigs are now one of the most expensive farm animals to feed. From the age of five months a pig will eat approximately one tonne of food per year, more if she is feeding piglets. By law, all pig food must be stored away from vermin, and, in truth, you don't want to feed squirrels and rats, or have them urinating over the feed, spreading disease causing organisms such as *Salmonella* or *Leptospira*, either.

Pigs must not be fed anything that has passed through a kitchen (domestic or commercial), even a vegetarian kitchen. That includes all foodstuffs including meat, old bread, peelings from the Sunday roast – you name it, and it's excluded.

The first outbreak of Foot and Mouth Disease in 2001 was started from a commercial farm illegally feeding their pigs untreated swill that was thought to contain imported meat. You can still feed fruit and vegetables that have been nowhere near a kitchen, e.g. your vegetable patch, and you can feed certain commercial bakery waste if the factory does not handle meat and has the appropriate Hazard Analysis and Critical Control Points (HACCP) procedures in place – provided you have registered with your local authority's Environmental Health department. You can also feed milk to your pigs if the cows/sheep/goats are on the same farm as your pigs, although I would be careful of feeding waste milk which is not fit for human consumption; it's considered unsuitable for human consumption for a reason, often because the animal has had antibiotics – I have a rule that if I wouldn't eat it, then I will not serve it to my pigs.

Recording of Medicines

All medication that requires a prescription from a veterinary surgeon or that is sold through a Suitably Qualified Person (SQP) at an agricultural store will need recording in a medicine book. If you can buy it off the shelf, then it doesn't need recording, e.g. antiseptic spray. Recorded details should include the pig's identification, the date/time, dosage amount, medicine batch numbers, and withdrawal times. You will need to refer to this book when completing the Food Chain Information (FCI) on your eAML2/ScotEID forms when your pigs go to the abattoir. There is now an option to use an electronic medicine book produced by the AHDB Pork called eMB. The website is https:/emb-pigs.ahdb.org.uk.

TIP

Write it in straight away or you will forget.

REGULATIONS IN THE USA

Agricultural Land Zones

Across the USA, land is zoned into different types: residential, semi-urban (residential and agricultural) and rural and open zone. Under individual state criteria, some hooved species are allowed in residential and semi-urban zones, but it is likely that you will need to live in a rural and open zone to keep pigs – even non-commercially. Smaller farms (e.g. less than three acres) may require restrictions on the number of animals over four months of age. To find out what type of zoned area you live in, check on your local state government website.

Premises Identification Number

Premises Identification Numbers (PID or PIN) enable complete livestock traceability and they form the basis of plans for the control, and prevention of agricultural diseases spreading. They are also used as an early warning system to notify animal owners of a natural disaster such as a flood or fire that could affect their animals. Premises where livestock are kept require PIN registration. Multiple PIN numbers can be assigned to one account if multiple land parcels are owned and used. If you will be using someone else's land to keep your pigs, you will need to know their PIN.

A PIN number is a seven-character unique identifier, consisting of a combination of numbers and letters, which is associated with a specific land location. Each state begins with a different letter, e.g. Alberta PIN numbers start with the letter 'A'.

PIN numbers are required or requested on many transportation documents when travelling your pigs. PIN numbers are also required when buying medications at a licensed retail outlet or when selling animals at an auction market. Agriculture programmes and grants may also request your PIN number as part of their eligibility requirements.

Movement Requirements

The movement of pigs interstate to another farm, to a market or to an abattoir requires a movement report signed by a state-accredited veterinary surgeon who has inspected and passed the pig herd/farm as free of disease in the previous 30 days. The movement report includes the name, address and PIN number of the producer, the PIN number of the destination farm, movement date, number, age and type of pigs, individual (farm, exhibition or market) identification or group* (abattoir) identification, named veterinary

---

* Individual ID is not required for the abattoir if all the pigs were born and raised from the same farm and have never been mixed with pigs from another farm.

surgeon, plus herd health status. The movement must be notified to the destination state's animal health official within three days.

Pig Identification

Pigs moving within state do not need identification, with the exception of cull-breeding animals, where the interstate movement rules apply. Suitable identification of pigs for interstate movement to a farm, market, exhibition, show or abattoir includes:

1. USDA '840' ear tags, aka Animal Identification Number (AIN) ear tags – suitable for any pigs except cull-breeding stock entering harvest channels (abattoir). They show the US shield emblem and an individual pig ID number beginning with the numbers 840.
2. USDA PIN ear tags – official ID for cull-breeding stock entering harvest channels and must show the US shield emblem and the PIN. They may be used as management tags with or without AIN for non-cull-breeding stock if wished. It is illegal to remove a USDA official ear tag and this is clearly printed on every tag.
3. Back tags – temporary ID label glued to the back of the pig and for the abattoir only on slaughter/feeder pigs.
4. A tattoo with USDA serial number assigned by state animal health official or state veterinary surgeon – for slaughter/feeder pigs.
5. Ear notching if recorded in an official pure-bred registry.
6. Ear/inner flank tattoo if recorded by pure-bred registry.

Records must be kept for two years and made available for inspection upon request.

Serial numbers of USDA PIN tags, back tags and tattoos are supplied by the individual state's animal health official, state veterinary surgeon or state/federal representative. You need to provide the serial numbers to an approved tag supplier for printing. Approved suppliers can be found at www.pork.org.

Movement of cull-breeding stock (sows and boars) from farm to an abattoir located within-state and interstate requires an official USDA Swine PIN ear tag. The tag must include a printed official US shield, the final farm's PIN number, state postal code of the PIN and notice stating 'UNLAWFUL TO REMOVE'.

Feeding Pigs

Feeding food waste to livestock is also known as 'garbage feeding' in the USA. It refers to only edible waste and is prohibited in 22 states, as shown in Figure 2.2, including; Alabama, Delaware, Georgia, Idaho, Illinois, Indiana, Iowa, Kansas, Kentucky, Louisiana, Michigan, Mississippi, Nebraska, New York, North Dakota, Oregon, South Carolina, South

Swine Health Protection

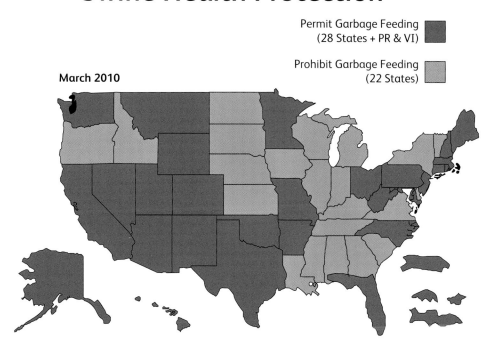

Permit Garbage Feeding
(28 States + PR & VI)

Prohibit Garbage Feeding
(22 States)

March 2010

FIGURE 2.2 A USDA map showing the states permitting and prohibiting garbage feeding.

Dakota, Tennessee, Vermont, Virginia and Wisconsin. In the remaining states where feeding garbage is permitted, it must be heat treated to at least 212° Fahrenheit for 30 minutes by a licensed operative at a licensed premises, and the supplied farms are required to have a state-issued permit: Class A permits where there is heat-treated meat or a risk of meat in the garbage, and Class B when it contains no meat and may also be known as 'exempt' garbage. Class B garbage does not need to be heat treated. Food waste that contains meat or has been exposed to meat has the potential to carry infectious disease, so, understandably, farms that feed Class A garbage are regularly inspected by state inspectors.

Federal rules state that farm household food garbage is exempt from these rules, but unless all the food waste originated from the farm, you could fall foul of the exemption.

NOTE: You can only export pigs outside of the USA if they have never been fed garbage.

VETERINARY SERVICES IN THE UK AND USA

It is recommended that you register with your local livestock veterinary surgery; some even do a free first visit to get to know your farm. This is not a rule or regulation but it is at least worth noting the telephone number down, so you're not scrabbling around trying to find it in an emergency. There are not many specialist pig vets around who are not tied to a commercial farm and there is no point ringing the local pet vet who treats your hamster!

In the UK, on the eAML2/ScotEID registration they ask for the name of your veterinary surgeon, so if you have registered, you can fill in that box as well. In the USA, a state-accredited veterinary surgeon will be involved in your interstate movement documentation.

GET YOUR HUSBANDRY RIGHT

By law, you have to look after your animals properly! Seems obvious, but the only way 'officials' can monitor how well an animal is being looked after, and prosecute if they're not, is to have laws passed telling you the requirements. In the UK, the Farm Animal Welfare Council (FAWC) reviews farm animal welfare, advises the UK government and governments outside the UK on legislative requirements and recommends the following 'Five Freedoms' for farmed livestock. You will see these written in different formats across many developed countries' rules and regulations, including in the USA. Wherever possible, strive to exceed them.

- Freedom from hunger and thirst – proper feeding regime and fresh, potable water
- Freedom from discomfort – good husbandry, shade, wallows, bedding, etc.
- Freedom from pain, injury or disease – preventative and prompt veterinary treatment
- Freedom to express normal behaviour – company of other pigs and, in my opinion, daily access to outside areas
- Freedom from fear and distress – kind, respectful handling

These are minimum, commonsense standards of care. It makes so much sense for both the pig and your pocket to keep your animals as healthy as possible. If you get the husbandry right and your pigs are in good health, then any bug, no matter how pathogenic or virulent, has less chance of taking hold.

What you are trying to achieve here is keeping your pigs secure, happy and healthy. Pigs will not care one jot about all the legal paperwork requirements you must fulfil, but they will thrive if you give them the correct environment. Good welfare can be achieved through a high standard of stockmanship, effective management, adequate housing and well-maintained equipment. I believe that pigs should always be raised outdoors in

TIP

Buy a powerful and rechargeable torch for those dark mornings and evenings. You cannot check your pigs if you cannot see them. The torch app on a mobile phone isn't powerful enough, I can assure you. I've tried.

as natural surroundings as possible for a happy life. They can, of course, be kept indoors their entire lives, but for me that takes some happiness away from them.

By law, you have to check and feed your pigs once a day, but, in my opinion, this is not enough and at least twice a day should be the norm. As a Commuter Pig Keeper, in the winter months you are most likely to be attending the animals once or even twice a day in the dark during the week. You need to be able to see your animals clearly to check all is well. You can either floodlight the whole area or invest in a rechargeable, powerful torch, and get into the habit of putting it back on to recharge after every use.

HOUSING

Pigs need shelter from the sun, rain, strong winds and the cold, so you need housing that is spacious enough, warm enough, sited in the shade and draught free. An ark correctly positioned, with sufficient bedding, can provide all of these. You do not need to buy an all-singing, all-dancing, expensive ark unless you choose to; pigs will not care as long as they are warm and cosy. You have a number of options.

Straw Bale Houses

The simplest design is probably the straw ark, constructed of some posts driven into the ground with straw bales lashed to them around three sides, with a corrugated sheet, wooden ply or tarpaulin roof – all held together with the robust and generous use of spare baler twine. Pigs like to destroy. It's what they do, so accept it as a challenge, and make sure the house is securely made to survive the 'pig test'. Using the posts will stop the bales moving during the customary scratching sessions, and the number you use will depend upon the size of shelter you need.

Homemade Pig Arks

There are no rules to say arks have to be curved – square or apex-roofed houses are equally good – but the apex roof will need a bigger base size to house the same number of pigs. This is because the roof is straight from the apex to the edge of the floor, which makes it harder for the pigs to turn around inside. Plywood can be used, even old half-round fencing rails recycled, banged to posts with a waterproof roof and lined with some hardboard inside to stop draughts.

Plastic Arks

If you have sufficient resources, you can buy plastic arks, which are made by several companies and are easily found on the Internet. They are lighter to move – by you and the pigs, it has to be said – but they can make a good mobile solution. They have an added advantage of being super easy to clean out and disinfect, but they are not so easy to repair if the pigs decide to take big chunks out of them.

Traditional Metal, Plastic or Plywood, Plus Metal Roof

These are possibly the easiest to source and needn't be horrific in price (Figures 3.1 and 3.2). Some piggy people say they are cold in the winter and hot in the summer, and that wooden apex arks are better, but we find that if you pack them with straw in the winter and provide shade in the summer, they are no problem. If you wish to spend a little more, you can buy insulated arks, which are cool in the summer and warmer in the winter. We farrow outside, all year round, in ordinary arks, without the use of heat lamps, without problem, so before you start throwing cash about, give it some thought. Remember, I'm in the UK and the extremes of weather seen in some states in the USA and the heat intensity seen in hot climates just don't happen.

FIGURE 3.1 A traditional wooden ark with a corrugated tin roof.

FIGURE 3.2 A plastic ark with a metal roof and a floor.

Do You Need a Floor?

Choosing an ark with or without flooring is a personal choice and may depend upon your local environment. On this topic my husband and I differ in opinion. I think if you live in a wet area in the winter, e.g. heavy clay, then flooring is a good idea. He disagrees and says that if you keep packing the ark with straw, a waterproof floor is maintained, you still have the convenience of moving the whole ark when a full clean-out is needed, and you can set fire to the old bedding in situ. I prefer to clean out the ark and disinfect between uses and think he is just finding excuses for a bonfire! If you can, choose a detachable floor that is secured in place with hinge-type bolts you can still move around relatively easily. You have to move floored arks, especially if you have heavy pigs as in wet conditions, as the ark can sink into the wet mud, causing water to seep between the floorboards. Of course if you live at the top of a sandy hill, or are only going to keep weaners in the summer months, then save your cash. If you are wise, you will choose an ark from a company that can supply you with a floor at a later date if you change your mind.

The Size of Ark Required

The size of the ark you require depends entirely on your requirements. Having smaller arks in a greater quantity, although more expensive, will allow for a greater flexibility, e.g. if you need to segregate one pig due to lameness or to segregate the sexes due to a delay getting them to the abattoir. If you are going to be introducing pigs over a time interval (known as having a 'rolling stock'), then penning pigs next to their future companions in a separate pen for a few days helps reduce the squabbles, and if you then just take down the dividing fence it also gives an additional ark for the new pig(s) to sleep in until they are allowed in the main ark with the others. This also works well when introducing older pigs to each other. If you pen them next to each other for a week or two, and feed them close together either side of the fence line so they get used to seeing and smelling each other, the squabbles are significantly reduced.

You might think that buying a massive ark in case you get a larger number of pigs in future batches makes sound economic sense. However, you then run the risk of the first small batches of piglets using a corner of the big ark as a toilet. It will be the furthest corner, by the way! Then the toilet smell is firmly in that corner even when you disinfect in between batches, and subsequent piglets will also follow suit. If you are stubborn and absolutely insist that you want a massive ark, or you are using an existing outbuilding or stable as their bedroom, then pack it with solid bales of straw to minimise the floor size available. Pigs will not foul in what they consider to be their sleeping quarters, so keep

TIP

Place the solid bales of straw with the bale twine down so they are less likely to spot it and open every bale for their amusement . . . daily.

TABLE 3.1 The number of different-aged pigs and appropriate ark sizes.

NUMBER AND SIZE OF PIGS	FLOOR SIZE (FEET)	FLOOR SIZE (METRES)
Two to four weaners to pork weight Two to three weaners to bacon weight	8 × 4	2.4 × 1.2
Two to three adult sows (not farrowing) One farrowing gilt/modest-sized farrowing sow One boar plus one wife Four to six weaners to pork weight Four weaners to bacon weight	8 × 6	2.4 × 1.8
One large farrowing adult sow One boar plus up to three wives	8 × 8	2.4 × 2.4
Five to six adult sows (not farrowing)	8 × 10	2.4 × 3.0

it small and increase the amount of size available as they get bigger. Why give yourself daily mucking out as an additional ten-minute chore?

So you have taken my advice and bought two arks to house your weaners, and then watch in dismay as they all pack themselves solidly into one ark. It is remarkable how many pigs will wedge into one ark, even if two are provided, especially in the winter to keep warm. In the summer they need a little more space for a cooling airflow, although if they have outside shade they often choose to sleep *al fresco*. Don't think I am wrong, as it is always useful to have the second ark in case of illness or a sudden impulse to buy another two pigs.

Positioning of the Ark

Arks should be positioned so the open door is not in direct line of streaming sun, driving rain or cold winds. A shady spot in the pen is an ideal place to position an ark. If at all possible, locate it on flat ground – somewhere the least likely to get waterlogged. Having said that, remember you will need to top up the ark's bedding, check why your lazy porcine hasn't got up when you are calling it, check on farrowing sows, etc., so try not to pick a spot that means you have a major hike to reach it, which in the winter may be swimming in thick mud. If there has to be a choice of what is best for the pigs and what is best for you, then sorry, the pigs win and you will have to arrive earlier to accommodate the extra time.

In the Northern Hemisphere, the ideal positioning is west facing because the wind is usually from warmer climates. North facing will allow the coldest winds and driving rain to enter the ark, south facing attracts the sun for most of the day. East facing, although better than

TIP

I personally would choose 8 x 6 ft arks as the smallest size to buy, as they give you the greatest amount of flexibility and are still reasonably light to move around. Most ark builders still quote sizes in imperial measurements.

north and south, may get the early morning sun. Unless you have purchased insulated or wooden arks, always provide additional shade or it could potentially be dangerous for the pigs. Vice versa for the Southern Hemisphere I suppose makes sense.

The indoor/outdoor mix system – whereby you make use of your existing buildings, such as a stable or barn, and give the pigs access to the outside during the day – is also an acceptable alternative. Here shade and shelter in the paddock from the rain becomes more important, as there is not an ark to retreat to. Pigs soon learn the route to their pen and will happily make their own way, often at speed when they are young, so have the gate open ready or they will scatter.

Bedding

Bedding needs to be warm and insulating, and like many other pig keepers, we use straw. Most pigs like to bed it down themselves, so just throw in the sections and the pigs will do the rest. They especially like to find the odd wheat head to munch on. Do not use hay as the primary bedding as it can hold the damp and lose its insulation properties. The odd section of hay, every now and again, is enjoyed to chew on. Sows also like to grab mouthfuls just before farrowing to weave into their nests, and we use it as advance notice of an imminent farrowing. Straw can be bought from agricultural merchants, direct from the farmer, or keep an eye out for adverts in the local papers. Always reject straw that is damp, has bird faeces on it, or has an unpleasant smell. Straw can harbour mycotoxin-producing organisms, so this last point is very important. We have also seen other forms of bedding such as shavings, peat and hemp used, and a friend of ours who rears pigs as close to naturally as possible in a massive woodland provides nothing, and the pigs bed down with leaves, twigs, grasses and plants that they find. To do this successfully, though, requires a high acre-to-pig ratio and access to enough raw materials.

Even if you are using a house within their pen, the pigs will still need somewhere within the enclosure to provide some shade from direct sunlight. This can be achieved by positioning your ark carefully, making use of natural shading such as trees and hedging, buying insulated arks, or providing artificial shading using posts, corrugated sheets or tarpaulins lashed together with spare baler twine. The shade should provide cover for at least the hottest points of the day, e.g. 11 am until 4 pm.

FENCING

What you are obviously aiming for is something that keeps the pigs where you want them, securely and safely. This is essential for the *Commuter Pig Keeper* as you cannot be called from work too often to catch pigs – so you need to invest some time and thought into this part of your husbandry. The only way we have found of (semi) guaranteeing this is to fence the required area with livestock fencing attached to wooden posts on the

FIGURE 3.3 *Stock fence and electric fence.*

outside of the post and then to line the inside of the pen with electric fencing (Figure 3.3). In addition to this, the whole of our land perimeter is post, rail and stock fence, so there is a double layer of fencing. This way, if the pigs do get out of the stock plus electric fencing, there is another layer of fencing before the big, wide world. I say 'semi-guarantee' as there are always occasions when the battery runs out, the fencing shorts out, etc. and the pigs are so inclined they can escape. Usually for us, this happens when we have some time off work and the 'do it tomorrow' gene within us is activated.

Once the pigs are trained to respect electric fencing, this alone will generally keep them in, but for pigs that are not used to electric, high-tensile stock fence is also required. The instinct of pigs when startled is usually to run forwards, especially the lop-eared breeds, whose field of vision is forwards and down. When zapped by the fence for the first time (and if a bit stupid second and third time), they will run forwards through the electric fence and out. This is also true when introducing new pigs – if they are pushed to the fence line in the scuffle that ensues, they will push through or even jump the electric fence. The stock fence stops this happening. Gateways can be electric with insulated handles (Figure 3.4) so you don't have to turn the fencing on and off all the time. We cover the gateways with either hurdles or a proper gate instead of the stock fence to make it easier to get the pigs out when required.

TIP

Make sure the livestock fence is the right way up. The smaller holes go at the bottom, which helps stop the smaller piglets getting out.

FIGURE 3.4 *An insulated electric fence handle.*

Some people have one smallish pen that is stock and electric fenced which they use to train pigs to the electric fencing, so that when they put them in their main enclosure they can just use electric and save a fair amount by not stock fencing everywhere. Some breeds, e.g. Tamworth and Mangalitza, always require stock fencing and electric together or you might as well not bother for the notice they take. However, the Duroc breed, I have been told, rarely challenge fencing and so livestock fencing alone would suffice.

We decided we wanted the extra security and stock fenced all the pens for peace of mind, and because we have more than one boar, we wanted to eliminate the possibility of them getting in together. I would recommend our belt-and-braces system to the *Commuter Pig Keeper*.

Electric fencing can either be run off the mains, solar power or via a leisure battery. The length of time a battery lasts depends upon the size of your battery, the length of the circuit, the number of strands used, how many opportunities the pigs have of shorting it out with mud and dirt, and your efficiency in keeping the weeds down under the fence line. Most units indicate when the battery is running low, but on a not-too-long run (200–400 m in total), with good weed and mud control, and a strong battery, you should get a good few weeks' continuous use before needing to recharge.

Mains power stops the need for battery recharging but it also has its downside: every time there is a power cut, your pigs could get out. Also, it can still short out and be inactive, and if you don't put a switch in near the pigs every time, if you need to repair or dig out some wire from the mud, you have to go back to the power source or you will get zapped as well. You have to run the wire from the mains to the energiser, which generates the volts, in the first place. We haven't used solar charging units but friends tell me they are reasonable at keeping the battery charged up. We just have one battery on charge, a couple charged and a few in use at any one time.

We run electric pig wire or tape around the inside of each fenced pen, and we have the most success using the orange wire with nine electrical connectivity wires inside, and the least success with tape. Other breeders swear by the tape, saying the pigs can see it better. The only rule we have found is don't mix different types of wire or tape in your system or it won't work properly. Choose one and use it throughout. The electric wire is attached to the wooden posts – on the opposite side to the wire fence, so it is less likely to short out – using insulating nails or screws, which hold the wire away from the posts. Pens can also be temporarily sub-divided with push-in insulating posts and wire if required, but only if it's not the end of the world if the pigs cross it – which they occasionally do.

For very young piglets and new weaners, we place the wire/tape approximately 15 cm/6 in from the ground so that, when they start rooting, there is a fair chance they will touch the wire and get a short, sharp zap. This is raised as the piglets get older, as when it is very low to the ground older piglets and adults often cover it with mud and short out the system. We have a complementary ring of insulator nails at higher positions as well so we can just raise the lower strand as necessary in a short time frame (Figure 3.5).

FIGURE 3.5 The insulator nails to hold the electric fence.

The wire needs to be at around top of the ear height when the pig has its snout to the ground, and we have found that a 15 cm (6 in) increment in height suits our pigs. The wire is then attached to an electric fence energiser unit and earthing stake, both out of the pigs' reach, which generates volts but not current – hence no death caused (Figure 3.6). We always carry fence testers in our pockets at all times, and as we know how many volts our fence is operating at, we use them to help us pinpoint any areas of shorting out. One end of the fence tester is placed in the ground and the other on the electric wire, and a reading is obtained by a series of lights denoting the volts. There are other systems that you can leave hanging on the electric fence which light up or beep when working, but a word of caution: pigs are clever and seem to catch on to this system as well. We like to see 6000 volts plus before we go to work!

Another solution, especially for temporary pens, is to use plastic push-in fence stakes (Figure 3.7), which have graded hooks at different heights up the stake to put the wire through.

The downside of using electric fencing is that the pigs become used to where the fence 'bites', so when you wish to move them, they are reluctant to cross the line. You can trick them through it by covering

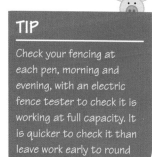

TIP

Check your fencing at each pen, morning and evening, with an electric fence tester to check it is working at full capacity. It is quicker to check it than leave work early to round up pigs.

FIGURE 3.6 *The energiser, battery and attachment to fence. The earthing stake in this energiser is the mounting stake.*

FIGURE 3.7 *Plastic push-in electric fence stakes.*

the line where the fence was with straw or dirt, or throw down a handful of pig food, so they slowly forage through and don't realise they have crossed the old fence line. Other methods include reversing a trailer into the pen and to move them, or building a triangle of hurdles around the pigs and slowly moving the hurdles across the line of the gate. It is worth turning off the electric fence when asking pigs to cross the line, as when it is on the floor it clicks loudly and the pigs think they will get a zap. Never try to physically force the pig over the line. You just won't win, you will just stress out yourself and the pigs – so what's the point? Our show pigs that are used to regularly crossing the fence line do so without fear, and care must be taken not to destroy that trust by accidently letting them touch a live fence. We always turn it off to make sure.

Size of the Enclosure

The pigs' pen needn't be acres of rolling hills – it could be an old vegetable patch you wish to turn over, or the end portion of your garden. In some ways, a series of smaller pens is nicer, as the pigs can be regularly moved and so have access to continuous greens. We like the idea that the pigs can break into a run when they have their mad half hours, making excited barking noises, and is always fun to watch. A space for a wallow is essential, and if you don't make them one, they will use the water from their trough. A pen size we often see quoted is 16 m x 16 m (52 ft x 52 ft), and this seems a reasonable size to take up to four (ish) weaners through to slaughter weight. We have some pens that are much larger and some that are smaller. The smaller ones tend to be used for farrowing sows, so they are only in them for six to eight weeks, but when we have a few too many pigs, we do use them without detriment for smaller pigs.

If you have any woodland, it can be used to create pig heaven. Don't bother clearing it – the pigs will do that for you. If you are using electric wire, you will only need to clear the perimeter before installing, then go round it once a week in the summer, and once every few weeks in the winter, with your strimmer, so that vegetation doesn't short out your system.

In the summer, our pigs hardly leave the woodland, and choose to clear a patch in the woods to make a bed at night time. Don't worry about poisonous plants and bulbs too much, as if pigs are well fed, then they are unlikely to eat them. However, if you see masses of poisonous varieties, then feel free to clear them. A few plants are instantly lethal to pigs; these include yew, monkshood, foxglove and hemlock. Others are accumulative in their toxicity over time, e.g. bracken and ragwort, so the damage can be insidious by nature.

WATER – THE MOST ESSENTIAL NUTRIENT FOR LIFE!

Pigs require water to meet their physiological needs. These include most metabolic functions, the movement of nutrients through to the body tissues and organs, lubrication of the joints, adjustments in body temperature, removal of waste, milk production, and for growth and reproduction – including semen production. A pig can lose most of its fat and half its body protein without dying and up to 10 % without affecting performance, but if it loses as little as 10 % of its water content, death quickly follows. Water balance is constant within the pig, with the most important source being available via drinking, although some additional water is manufactured in the body via the breakdown of dietary carbohydrate, fat and protein. The pig loses water via urine, faeces, respiration and from the skin. This intake–output balance is directly affected by a multitude of factors, including their age, what they are fed, weather conditions, health status and type of husbandry, so the ideal quantity required cannot be quoted in absolute litres.

Studies on indoor-reared commercial pigs – usually performed on ones kept in constant thermo-regulated environments – have shown such a wide range in 'average' amounts drunk at different ages. It is impossible to say how much credence can be given to the figures. Pigs fed on a dry feed have been reported to drink between 1.5 and 3 litres/quarts per kg (2.2 lb) of food consumed, and where quantities have been published by different researchers, these have ranged from 1.9 to 2.8 litres/quarts per day each in newly weaned pigs; growing pigs 5.5 to 20 litres/quarts; in-pig gilts and sows 11 to 25 litres/quarts; working boars 20 litres/quarts, and lactating sows 10 to 30 litres/quarts. Our outdoor-reared traditional pigs don't seem to drink anywhere near the higher quantities, and it is highly likely that not only are they fed less dry food per day, but they are getting moisture from the ground when rooting and drinking from puddles and wallows. We do notice an increase in water consumption when it is hot and during lactation, and when flushing the gilts and sows prior to service, as the amount of dry feed increases. Maybe the commercial pigs have a higher level of water consumption because they are faster growing and so fed correspondingly higher amounts and are fed higher protein levels when growing. The amount of water consumed increases proportionally with the amount of crude protein in the diet, so pigs fed on 13 % crude protein will drink less than those on 16 % or 18 %. Also pigs fed a pelleted feed require more water than those fed a wet meal, and also, perhaps surprisingly, a dry meal. It is reported that underfed pigs increase water consumption, particularly in gestating sows as they presumably attempt to 'feel full'. Pigs fed with a diet supplemented with permitted fruit and vegetable waste will also drink less.

Pigs affected with diseases require more water than healthy pigs of the same age and body weight, especially if the illness is accompanied by diarrhoea or the animal has a high temperature caused by a fever. This makes perfect sense as more water is being lost. If required, add an additional trough near the sleeping quarters to encourage drinking; if they have to trek across a long distance or through thick mud, they might not bother. This is applicable for new healthy weaners in an unfamiliar pen as well. Maximum fresh-water availability during lactation can also help to avoid urinary tract infections.

Don't forget the boars. When working, they perform with considerable physical exertion and produce up to 450 ml (15 fl. oz.) of semen at each service, all leading to a water deficit that needs replacing.

Ideally all water provided should be fresh and clean, but, unfortunately, within seconds of cleaning out and refilling, pigs will put mud and debris back in the water container. You have to do your best and clean out troughs at least twice a week and top up twice a day – more if required. Never allow water to go stagnant and position the troughs away from posts/solid fences so when birds perch on the post they don't defecate into the trough. A large nail sticking out on the top of the post can help prevent it. While a pig will drink quite happily when there is mud in its water, if the water is contaminated with faeces or bird droppings, mouldy vegetation or green slime, it may not.

You should use water troughs of different sizes for different ages, and always be mindful that piglets, which need water provided from two weeks of age, may be able to get in but not out, so choose trough depths carefully (Figure 3.8). For our weaners, we use small feed troughs, which, although they continually tip them over for fun, do not pose a drowning risk. To make sure they do not run out, during times we are not there, we provide at least three troughs and semi-bury one of them so it is harder to tip over. We have had success with using the white bowls that are designed to fit inside tyres for the slightly older piglets as they can't seem to gain purchase on the bowl part with their snouts and so, apart from when they sit in them, they remain available to drink from. For the adults we use deeper water troughs, which, where possible, serve two pens, with half in each. If you use electric fence to divide the pens, make sure that the wire cannot touch the trough as the pig will not drink if it gets zapped each time. These are perfect for adults, as they are too heavy to tip up, but are not an option for piglets, as even if they could get in, they would not be able to get out and would very likely drown.

Be careful of relying on streams as a source of drinking water, as they may be contaminated from further upstream. Sheep, for instance, seem to like to die in or near a stream and, in some instances, the water authorities could ban you from using streams to provide water for livestock. Other alternatives are nipple drinkers, which are teats attached to the mains water supply via pipes, which do not require cleaning out and

FIGURE 3.8 Different-sized water troughs.

have continuous fresh water. Careful maintenance is required as they can get stuck and flood everything, and as the flow rates are known to affect the ability to digest feed efficiently, they can be detrimental if set incorrectly, particularly if there is too much competition for too few nipples. Small automatic fixed bowls are also popular. They generally stay cleaner than troughs and cannot be tipped up, but have to be very securely fixed with no pipes left exposed that might be chewed. Large troughs with automatic water filling are a nightmare to clean out thoroughly as you cannot tip them up and give them a power spray. You will find a large layer of mud at the bottom of troughs even after a week, especially in the winter as there is mud all over a pig's face when it thrusts into the trough. In the summer you have to remove the algae, which can turn the water green if you don't get rid of it. There are easily cleaned automatic filling troughs with a tipping mechanism, which look like the perfect compromise, and as soon as the cash flow allows, they will be on the shopping list.

Muddy wallows are an essential tool for the outdoor-reared pig to keep cool in the hot weather via transfer of their body heat to the cooler mud (Figure 3.9). They often dig at the wallow first with their snouts to gain access to the coolest mud and then slide in, rubbing every inch of themselves as they go. The mud covering also provides an excellent

FIGURE 3.9 A muddy wallow – no sunburn risk here.

FIGURE 3.10 *A trough with inaccessible water.*

sunscreen, which is essential in the lighter-coloured pigs. Pigs kept indoors will attempt to wallow in spilled water or other wet substrates, especially on warmer days if the barn isn't temperature controlled. Outdoors, they are more than happy to lie in a stream or water trough if big enough, so it makes sense to create the wallow in a suitable area – not near the gate, and in the shade if possible. If you don't provide one, the pig *will* make one using the only water they have available – from their trough! Pigs are clever, but they don't have the foresight to see that this reduces the water available for drinking.

Hot weather puts additional pressures on the pig to keep cool, and due to a deficiency of functional sweat glands in a pig, expulsion of heat via urination is one way they can cool down. However, this does conversely require them to take in more water. Pigs are instinctively clever enough to limit feed intake in hot weather to reduce the metabolic heat generated by digestion and it can really assist the pig to maintain its water balance by feeding at the cooler ends of the day. Any outdoor pig keeper with a crow problem will probably be feeding at dawn and dusk anyway!

In cold weather, if you suspect the water troughs (Figure 3.10), pipes or hoses will freeze up, have some form of contingency, such as draining the hose after use and filling water containers outside each pen every time, so that if the tap is frozen in the morning, you can top up the troughs using a bucket.

If you are not going to be around for a few hours in very hot weather, arrange for someone to come in and check and top up the water. We have a neighbour who will check the water levels during the day during very hot spells and top up if required. It is quite reassuring when she confirms that they are all fine and didn't need any extra water, as we provided enough troughs and a decent-sized wallow. Water deprivation can quickly cause death in pigs (see Chapters Five and Nine).

Given the correct husbandry conditions, and with plenty of fresh palatable water, pigs are pretty good at looking after themselves. How proficient you are at providing their needs isn't usually challenged on a day-to-day basis; however, the difference between good and excellent water management lies in how your system provides for the needs of any that become unwell or are vulnerable in other ways.

NOTE: In the UK, you will have to fit a non-return valve to automatic water containers and may be required to inform your water authority when you fit a new one.

BODY CONDITION SCORING

We judge the condition of our pigs using body condition scoring (BCS) rather than just looking, and if the body is scored as too thin, the feed is increased; if the pig is too fat, it is decreased. This may require penning pigs of a similar constitution together. BCS is a method of judging the overall condition of a pig and allocating a score between 1 and 5, with 1 being malnourished and 5 being grossly obese (Table 3.2). What you are aiming for is a body score of 3 to be maintained at all times through the life of the sow, no matter what stage in the breeding cycle. If the sow starts dropping in weight during lactating, increase her feed intake; if she cannot eat enough of the sow feed at 13 to 16% protein levels, help her out by feeding a higher-energy feed such as growers pellets – but do not feed creep pellets as the levels of copper may be toxic. Conversely, if she is too fat, reduce her feed accordingly. Body score regularly when you are giving your pigs a fuss so you know at least on a fortnightly basis the body score of each pig. This way, any changes you have to make to their diet need not be too drastic and the sow will barely notice. We always divide the daily feed into morning and evening rations, so we automatically get to check their general health, water and fencing twice a day. Unless the pig is on its own for quarantine reasons or is just about to farrow, we feed over a wide area so that even the less confident pigs get enough to eat. If you have problems with birds stealing the feed, then you can feed in piles far apart, but a pig that eats quicker than another may steal the remains of the other pig's feed and increase its intake.

TABLE 3.2 How to body condition score pigs.

SCORE	VISUAL		BY TOUCH
1	Emaciated		Backbone, shoulders, hips and ribs highly visible and defined. Be careful when touching these sows as it may hurt
2	Thin		Backbone, shoulders, hips and ribs are noticeable and easily felt with no palm pressure
2.5	Moderately thin		Backbone, shoulders, hips and ribs can be felt without palm pressure
3	Perfect		Backbone, shoulders, ribs and hips can be felt with gentle palm pressure
3.5	Slightly overweight		Backbone, shoulders, ribs and hip bones can be felt with moderate palm pressure
4	Overweight		Backbone, shoulders, ribs and hips cannot be felt with moderate palm pressure
5	Obese		Backbone, shoulders, ribs and hip bones cannot be felt with heavy palm pressure

If necessary, we put the two greediest, or the two lowest-ranking, pigs in separate pens, rather than worrying and watching them get thinner.

FEEDING

As mentioned earlier, there are many UK and USA regulations regarding the feeding of pigs which can be downloaded from the www.gov.uk or USDA websites. You may purchase, pick or grow fruit and vegetables and feed to the pigs directly as part of a mixed diet, but with caution, as I will explain. This is not something we do; the pigs get their full diet as proprietary pig food, and we only use fruit and vegetables as the odd treat.

Commercial Feeds

The easiest and most accurate way to feed pigs is using a pre-made, commercially available pig feed. There are a number of manufacturers that typically stock feed in 20 kg or 25 kg sacks (UK) and in 25 lb, 40 lb and 50 lb bags (USA). As with most animal feeds, buying in larger quantities or bulk saves you money. Just check the best-before dates before buying in a lot, as you need to know if you can use it all before that date and, by law, you will have to store it away from vermin.

Different feeds are available for varying ages and stages of development, and even just for rare breed pigs. A piglet will eat up to 2.5 % of its body weight per day, and from birth to eight weeks will put on 10–11 times its birth weight. Feed comes in a variety of formulations and may be called by the size of the preparation, e.g. pellets, nuts, rolls, pencils, and then in age names such as creep, weaner, grower, finisher and sow (Table 3.3). Correct feeding early in life is critical: muscle grows exponentially, so the more quality muscle the piglet can put down early in life, the better the feed conversion ratios later in life. A good reason to buy from a reputable source! There are also feeds in the USA specifically for show pigs. These are the same feed as commercial feeds but with

TABLE 3.3 Different types of proprietary pig feed pellet sizes.

NAME	PELLET SIZE (DIAMETER)	
Roll	15 mm	0.6 in
Nut	5 mm	0.2 in
Pencil	3 mm	0.12 in
Pellet/creep	2 mm	0.08 in
Meal (powder)	N/A	

The smaller the pellet, the greater the surface area by weight available to the digestive process of the pig, but also the greater the chance of getting lost in the mud or bedding.

antibiotics added, usually Lincomycin or Tylosin. The use of them prophylactically to prevent 'catching something' at a show should be frowned upon as it will no doubt contribute significantly to the growing worldwide anti-microbial resistance problem. There is also the risk they will be used as growth promoters. Luckily they are as expensive as creep feed, which may put most people off using it. Chapter Eleven lists the current (2016) feed prices.

You would have thought that by keeping to the correct feed for age, and following the manufacturer's feeding instructions, you would know that your pigs are getting the correct balance of nutrients for growth, breeding or maintenance. However, the higher-protein grower and finisher pellets have been designed with modern, commercial pigs in mind. Commercial breeds are much faster growing than traditional and heritage breeds and are ready for slaughter at 16 to 18 weeks. If these are fed in the recommended quantities to your traditional or heritage breed, it is likely to run to fat, so save the extra money and go for sow and weaner pellets/rolls.

If you have chosen a faster-growing modern breed of pig, then the use of the commercial preparations will be cost effective and of benefit to the pig and the final meat. With the exception of very young piglets, whom we introduce to growers pellets or creep feed from seven to ten days old, we feed sow rolls from the beginning to the end, as do a lot of traditional breed keepers. Some traditional breeders use rolls in the winter as they take a longer time to disintegrate in the mud and so can be found easily by the pigs, and then use nuts in the summer. We use rolls all year round for ease of ordering. Some traditional breeders never use creep feed at all and the piglets just pick at Mum's nuts or rolls. We prefer to start them off on creep as it helps lay down that early muscle which can then be built upon as the piglets get older; and it helps occupy the skittish mind of a piglet and can prevent behavioural problems such as biting the sow's teats or their siblings' ears and tails. The creep feed and sow rolls are then mixed in differing ratios as they get older, so by the time they are eight weeks old they are eating only the sow rolls.

We follow a standard feeding protocol of 1 lb (0.45 kg) of food per day, per month of age up to four months for fattening stock and five months for breeding/showing stock. So when you get your weaners at eight weeks old they will be on 2 lb (0.9 kg) of food per day (1 lb am and 1 lb pm). That is increased by 1 lb (0.45 kg) per pig per day, so: 3 lbs (1.35 kg) at three months, 4 lbs (1.8 kg) at four months, and, if they are to be kept as breeding stock, 5 lbs (2.25 kg) at five months. We do not accurately weigh every pig's food every day, but we have a scoop which we know holds a certain weight and then just count out the number of scoops per day. Other breeders are known to *ad lib* feed to appetite until 12 weeks of age and then restrict to 4 lb or 5 lb per day. I did try this method, but without specialised feeders protecting the feed, I got an awful amount of wastage and I worried about attracting

TIP

Use imperial weights when measuring feed for the pigs as it makes it easier to remember: 2 lb per day at two months, 3 lb at three months, 4 lb at four months and 5 lb at five months.

vermin. At the end, the pig carcasses finished the same at six months of age as with my usual method.

Our standard feeding protocol is a starting point of what to feed, and we find with the British Saddlebacks and Middle Whites that it works very well in 95 % of our pigs. The Gloucestershire Old Spots we owned seemed to run a little fat at this level, so we penned them separately and fed slightly less per day, although on occasion we penned them in with other pigs that required a little more, and as they tended to be a bit slow in feeding, so they naturally took slightly less. Other pig breeds may eat much less than these amounts (e.g. the Berkshire and Kunekune), or more than these amounts (e.g. commercial modern breeds), so take advice from the breeder you purchase your pigs from. Don't just feed by weight per day – perform body condition scoring described earlier in this chapter so you can proactively monitor if your pigs are too thin or too fat and adjust the feed or pen mates accordingly.

Never allow uneaten food to be left, as in a short space of time it will have attracted vermin, flies and birds which may increase the amount of disease on your farm, compromising your bio-security measures. If they haven't eaten their ration within 30 minutes, it's likely you are feeding too much. If birds are a real problem and you have enough pigs to warrant the expense, there are now feed additives that repel birds, which can be used for short periods of time to move them on, without harming them or your pigs. The other option is to feed at dawn and dusk when the birds are roosting.

Explaining Protein Levels in Commercial Pig Feed

It is worth having a basic scientific understanding of protein levels in manufactured pig feed and how they can affect health, growth rates and profits. Proteins are required by the entire body to function and they work synergistically alongside other intrinsic nutrients such as carbohydrates, fats, vitamins, minerals and water, maintaining bodily functions. Each protein within the pig has a specific role, including muscle and organ development and maintenance, hormones and enzyme production, immune system function, foetal development and lactation, and transport and storage of nutrients around the body. They are continually utilised and require constant replacement.

Some proteins are termed the 'building blocks' of the body and are heavily involved in the construction, repair and maintenance of hair, skin, eyes, muscles (meat) and organs. This is why piglets, weaners and growers need more protein per kilogram of body weight than mature sows and boars; they are growing and continually developing new tissue.

There are a vast number of different proteins that have been identified, and they are all constructed from different combinations and ratios of just 22 amino acids. Of these, 10 cannot be synthesised by the pig and have to be supplied via the diet. They are, therefore, known as essential amino acids, and are arginine, histidine, leucine, isoleucine, lysine, methionine, phenylalanine, threonine, tryptophan and valine. The

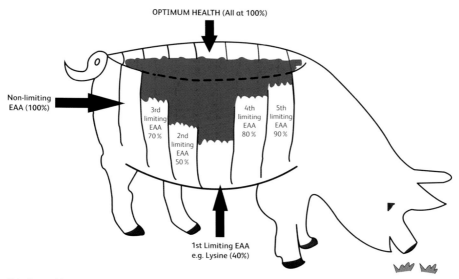

EAA = Essential Amino Acid % = Proportion supplied of optimum level required

FIGURE 3.11 Limiting essential amino acids.

remaining amino acids, although equally important for optimal bodily function, can be manufactured by the pig using other available dietary proteins and are less likely to be in short supply. When the supply of one of the essential amino acids does not meet the pig's requirement, it is said to be 'limiting', as it prevents the utilisation of the other amino acids. In pig diets, the most influential of the limiting essential amino acids is lysine, and, for this reason, most commercial feed formulations quote the requirement of the other essential amino acids as a ratio to lysine. A simplistic example would be: if the lysine is at 40 % and all the other nine amino acids are at 100 % of their individual nutritional requirements, then *all* the amino acids will be at 40 % of optimum availability to the pig (Figure 3.11). Once the limiting amino acid is added to the diet in sufficient quantity, the next amino acid will become limiting (second, third, fourth limiting, etc.).

Imagine the pigs' health as a barrel, with each of the staves representing the amount of supplied essential amino acid and the water level as optimum health. In this example the water in the barrel would spill out to the level of the lysine and the utilisation of the other amino acids would be below their actual level supplied. If the lysine was increased to 100 %, then the water would spill to the level of the second limiting amino acid, then the third and so forth.

The Effect of Limiting Amino Acids in Feed Formulation

The dietary lysine requirement for the early weaned piglet, i.e. less than four weeks old, is quite high (1.70 %), but this decreases to 1.2 % by six weeks of age and continues to steadily decrease to 0.9 % during late finishing. Adult diets optimally contain around 0.6 % lysine during gestation, with more required (1 %) during lactation. It should be noted that most published data is for the faster-growing modern breeds of pig and it is probable that the slower-growing traditional breeds would have optimum levels slightly lower than the percentages quoted, until adulthood.

The amino acids most likely to be at deficient levels are lysine, isoleucine, tryptophan, threonine and methionine. Different cereal crops have different amino acid deficiencies, a primary reason why blends of crops are used to produce balanced compound feeds. Corn is deficient in lysine and tryptophan, while barley and wheat are low in lysine and threonine, and the first limiting amino acid in soybean meal is methionine. In practice, lysine supplementation allows for the use of lower-cost crops, e.g. maize rather than soy, while maintaining high growth rates. Milk protein is well balanced in essential amino acids and highly digestible but too expensive to be used in all but creep feed diets.

The good news is that when feeding commercially available diets, all these issues have been calculated for you and the variability between the same crops from different sources taken into account. Manufacturers make feed for different life stages, but it is the pig keeper who must decide which feed is appropriate. As traditional breeds typically require less protein than the modern breeds the feed is usually formulated for, a certain amount of guesswork still remains for most small-scale or hobby keepers of traditional and heritage pigs.

Don't be tempted to feed high levels of protein, thinking it will solve the protein-balance problems, as a surplus of crude protein fed to pigs has been shown to be negative in pig production. Although protein is a utilisable energy source, if a pig consumes a sufficient intake of energy preferable nutrients such as carbohydrates, then any excess protein is available to be used to create fat. Additionally, studies have shown that one limiting essential amino acid will cause a pig to convert some of the remaining unutilised protein into lard instead of producing muscle. Feeding a perfectly balanced diet should mean less fat production and more meat production – it takes more calories of pig feed for the pig to create fat than to create muscle: 9 kcal for 1 g of fat and just 4 kcal to create 1 g of muscle!

It is also a waste of your money, as an over-supply of crude protein will result in inefficient use of energy to break down and excrete feed, meaning you are paying for calories solely used to break down excess protein into faeces.

There are other advantages to reducing crude protein in the diet. Pigs drink less water, naturally drinking between 10 % and 30 % less water compared with identical groups fed high-protein feed. This can have significant financial savings if you have a lot of pigs and are on a water meter. Also, it can mean less manure and urine to dispose of, and a reduction in the amounts entering your soil. This reduction in manure volume can have a knock-on effect if you are required to comply with Nitrate Vulnerable Zone regulations. In typical feeding practices, only approximately 30 % of protein contained in feed is utilised effectively for protein synthesis. The rest is excreted as nitrogen via manure, urine or gas. Measures taken which reduce the crude protein content in pig feed have been shown to reduce nitrogen emissions by an average of 10%. Faeces production generates many volatile, gaseous compounds that originate from excess amino acids in the diet – not a problem if you have no near neighbours, but any close by might appreciate the lower levels! Non-utilised crude protein is additionally available to be used by pathogenic bacteria, especially in the large intestine where it can impact on the absorption of water and cause intestinal disorders. If those reasons haven't convinced you, then the cost savings might – lower crude protein feeds are significantly cheaper to buy.

Reducing the crude protein levels wherever possible is the smart approach for both pig health and profit. With the right feed, fed at the correct levels, pig keepers can make an important contribution to optimised gut performance, the suppression of pathogens and supporting a strong immune system. And you will be helping the environment! We changed our feeding regime to a lower crude protein a few years ago, after much angst and lengthy discussions with a pig nutritionist, and have saved approximately £1500 a year in feed costs with our modest-sized herd and moved our herd health positively forwards. Give it some thought!

Alternative Feeding

We all would like to save money when feeding our pigs, and for some that means alternatively sourcing feedstuffs. Be careful, as some have shown to cause health issues in pigs. It pays to know what you are feeding is of benefit and how much is ideal. There is a general myth that pigs can eat anything and remain in perfect health, producing large litters of pigs and wonderful meat, to boot. People feeding 'on the cheap' often boast that their pigs have a natural diet as a part of their marketing – but there is nothing natural about feeding trailer loads of potatoes one week and a pile of bakery waste, broccoli and cabbage the next. Bizarrely, some supposedly healthy vegetables have been clinically proven to have a negative effect upon a pig's protein digestion capability and some can even cause clinical illness. Nutritional balance is the key.

If you wish to supplement your pig's diet with legally sourced feedstuffs, you need to be aware of Anti-Nutritional Factors (ANFs). These are substances that, when present in feed, reduce the availability of one or more nutrients. These can be substances that depress protein digestion or interfere with the utilisation of vitamins, e.g. mould mycotoxins, pesticides and plant toxins. They can also be substances generated by the pig from eating a natural product which can cause diminution of the digestive/metabolic processes, making the feeding of certain 'natural healthy' vegetables counterproductive at certain levels (Table 3.4). Where high proportions of vegetables (treats don't count) are included in diets, then supplementation with essential amino acids is

TABLE 3.4 *General guide to alternative feed sources.*

VEGETABLE / OTHER	CONTAIN ANFs	REPLACE 1 KG / 2.2 LBS OF COMPOUND FEED WITH. . .	MAXIMUM ADVISED DAILY REPLACEMENT LEVEL IN DIET	ADDITIONAL INFORMATION
Raw potato	YES	6 kg / 13 lb	20%	Green potatoes contain the toxin solanine and should not be fed to pigs
Cooked potato (UK not from a kitchen)	NO	4 kg / 9 lb	No advice found	Cooking enhances nutritional availability by 40% and inactivates ANFs
Brassicas (Cabbage/ sprouts/broccoli)	YES	2.6 kg / 6 lb (cooked cabbage)	< 10%	ANFs adversely affect metabolism and digestion
Carrots without green tops	NO	8 kg / 17 lb	No advice found	Large amounts can colour the carcass fat yellow
Swedes and turnips	YES	10 kg / 22 lb	20%	Contain pectin which can impair nutrient utilisation
Parsnips	YES	6 kg / 13 lb nutritionally but best avoided	0%	Significant health implications reported, e.g. increased photosensitivity causing ulceration of the mouth
Fodder beet	YES (oxalic acid in the leaves)	6 kg / 13 lb	50%	Feed with tops wilted or removed. Excellent replacement feed high in carbohydrates
Whole milk (from own farm in UK)	NO	4.5 kg / 10 lb	Personal choice	Limitation due to fat deposition in carcass, as high fat content in milk
Brewers grains	POSSIBLY	Analysis would be required to provide advice	20% sows/finishers 10% growers	Be careful of moulding, fermentation and toxin production if it is stored before feeding

KEY: AVOID FEEDING TO PIGS   FEED WITH CAUTION   A GOOD NUTRIENT SOURCE

necessary if there is to be no reduction on performance. However, carcass quality and kill-out percentages (KO %) may still be affected.

Another common misnomer is that pigs can eat vast quantities of acorns without consequence. In young piglets, belly ache and constipation are common after eating too many, so access should be restricted. The pigs you hear about that are fed solely on acorns on the European continent and in the USA probably have several generations of acorn-eating pigs in their ancestry and so have natural tolerances.

Fruit

There is scant scientific evidence on the quantities of fruit that can nutritionally replace conventional pig feed. Comparatively there is very little protein in fruit, with most of the calories/energy provided coming from their carbohydrate content, most often in the form of a sugar. Pigs have quite a sweet tooth, making fruit a popular treat, but too many carbs can lead to weight gain, so exercise some caution in quantities fed. I have read some anecdotal evidence of one third of the recommended maximum quantity of vegetables making an average of 6 kg fruit replacing 1 kg of compound feed, but only to a maximum of 5 % of the total diet. This is approximately 600 g, which is about four large apples – so more of a treat than a feed replacer.

What is important is the quality of the fruit that's fed. Fruit that is past its best rapidly grows a plethora of moulds, which can be harmful to health and have been known to cause sudden death in extreme cases. Fruit also starts to ferment once harvested, producing alcohol, and so if fed in large quantities you could end up with a drunken pig staggering about. This is particularly applicable to apple pomace (cider) and potato pulp (gin) by-products. It's quite something to watch a pig eat a bunch of grapes – spitting out a perfectly cleaned stalk after a quick swirl around their mouth or a stone out of a peach or nectarine.

Breaking the Pig Swill Ban

There is a drive by an environmental pressure group called The Pig Idea to resume feeding pigs human food waste – also known as 'swill'. Currently, it is illegal to feed this to pigs within the European Union, for sound biosecurity reasons, and it is illegal in many states in the USA as well.

The Pig Idea's aims are simple – to decrease the use of human food grade virgin crops such as wheat and soya in the feeding of pigs, decreasing the environmental impact of grain production whilst, at the same time, recycling out-of-date food waste into pig feed and reducing landfill. This, they say, would also reduce feed costs for the currently financially stretched pig industry, particularly the smaller-scale pig farmers producing pedigree and rare breed pork.

At first glance the idea could be considered a perfectly reasonable and practical solution to a growing environmental problem. So why has the National Pig Association (NPA), representing the commercial pig industry in the UK, been very clear in not supporting the idea as feasible, and the British Pig Association (BPA), representing the small-scale pedigree pig breeders, taking the same stance? In addition, why is the specialist smallholding press widely supporting the NPA and BPA in condemning the idea as unworkable at best and dangerous at worst?

Unfortunately in modern times simple solutions to any problem are hard to achieve. The ramifications that would and could be caused by The Pig Idea assists both the NPA and BPA in drawing their conclusions. Food in bygone times was uncomplicated. Our ancestors' diets contained unprocessed foods with low salt and sugar content; they ate less imported meat and no ready meals of questionable provenance; foods contained less non-nutritious bulking agents, less dubious ingredients such as artificial sweeteners. Yet, even in those times, swill was known to effectively transmit significant disease-causing organisms – and until it became illegal to feed untreated swill from the mid-1970s, it was linked to many diseases, some of which are currently eradicated from the UK and considered so dangerous they are notifiable by law, e.g. Swine Vesicular Disease, Foot and Mouth Disease, Classical Swine Fever. Porcine Epidemic Diarrhoea Virus is now (2016) notifiable in the UK following the decimation of the pig industry in the USA in 2013–14. The UK, following the 2001 Foot and Mouth Disease outbreak, knows what can happen when raw swill is fed to pigs on one, single farm. The impact of any of the eradicated diseases reappearing would be costly to pigs' welfare, to business and to the taxpayer. These diseases still exist in countries that we import from, up to 60% of the meat consumed in the UK.

The Pig Idea premise is that food waste not fit for human consumption should be collected from large manufacturers, retail outlets, caterers, smaller food businesses and individual households and taken to local or regional collection points where it would be processed and sterilised and sold to feed manufacturers to replace the grain content of pig and poultry feeds. They suggest this would lower costs to the pig industry, and particularly help the small-scale keepers. Even without any other consideration apart from cost, the reality is that the collection, sterilisation, quality control, turning the waste into a format usable by feed manufacturers, e.g. a powder, and the delivery to the manufacturers is going to be very high in terms of energy costs. As The Pig Idea points out, the industry is already burdened with high feed costs, and this approach is unlikely to alleviate the situation. The costs could (arguably) reduce in time, and so may potentially be achievable; however, there are numerous other considerations of varying implications that need to be taken into account.

There is also the issue of provenance. At a basic level, how will organic pig feeds be made and processed? Advocates of eating organically pay a premium to eat foods with minimal interference from chemicals in their production. Corporate industries around the world have been built upon that whole premise. Will there be separate collections for organic feed waste that are kept apart from non-organic waste, and if so how will this be policed?

How will the provenance of any food waste be collated? Presently all feedstuff used in the manufacture of pig feeds is fully traceable to farms or sources of origin. The Pig Idea has acknowledged that, every year, one million tonnes of human grade food industry by-products are already incorporated into pig feed where available. So incorporating suitable food waste into pig feed is not new. However, importantly, this type of waste, with identifiable ingredients, in known amounts with full traceability, is safe and cost effective to include. Those with smallholdings or a friendly grocer already legally utilise fruit and vegetable waste. This again is fully traceable and, importantly, carries disease risk. The retail industry is experiencing enormous problems with traceability and provenance at the moment – not least in terms of the horsemeat scandal and the recent suspicion that foreign meat is being sold as UK high-welfare meat. We are only too aware of this, due to the extensive efforts of the UK food industry governing boards. So how can a plethora of food waste from multiple sources ever have a provenance confidence? Quite simply it cannot.

How would the processed or treated swill be incorporated into pig feed manufacturing? To be in a usable format the waste would need to be converted into a powdered form immediately, or storage solutions would have to be found for food already at the end of or beyond its nutritional life. Manufacturing plants would need adapting, all of which would cost considerable amounts of money, pushing costs up. All food stored prior to processing would have to be kept away from flies and vermin, both of which are capable of spreading disease as a vector and/or host. Fly eggs hatch into maggots within a day in the correct environment, which a household or factory waste bin will provide even on a moderately warm day. No household waste storage can be monitored adequately before collection and putrefying food is not of suitable quality to feed pigs.

What about the possible contaminants in multiple-sourced food waste? If human food is being diverted from landfill and collected by households, it will be mixed in with many undesirable contaminants such as salt, artificial sweeteners, colourings, preservatives, antioxidants, pesticides from unwashed fruit and vegetables, detergents, bleach, discarded antibiotics, other human medicines, contraceptive pills, broken glass, pet food and non-edible rubbish. This will all be passed through into the feed and therefore the meat; these unseen contaminants cannot be removed. They will be in both solid and liquid forms, so unless every donation to the waste is checked by extensive and expensive analysis, they will all pass through undetected. If the processed food waste is batch tested, what happens if these contaminants are found in a feed batch . . . expensively produced landfill?

Food not fit for human consumption has a lower nutritional value, as nutrients rapidly deplete once harvested, so the feedstuff collected will already be past its best and well on its way to rotten.

Already substandard food, which is then stored, sterilised and processed, will be of a much lower nutritional content, lower even than raw swill. As it would be of variable quality and nutritional content, it would require considerable balancing to produce a diet that could be fed in known, manageable quantities on a daily basis. It is well documented that pork quality and taste declines when pigs are fed swill, but what about pig breeding programmes? With the unknown contaminants in the waste feed, sows and boars may have reduced fertility and increased susceptibility to disease – making our rare breeds even rarer. Fattening pigs will also take longer to reach the weights required, all pushing up the price of meat, compromising the very lure The Pig Idea is promoting to the pig industry.

What about waste from the abattoirs and meat-processing plants? Would these be incorporated into pig feed? There would also be meat in all the collected waste from households. While pigs are, indeed, omnivores, and the associated risks from feeding them meat is lower than in ruminants from Transmissible Spongiform Encephalopathies, there are numerous other viral, bacterial and parasitic disease-causing organisms which they are highly susceptible to. The complete removal of pork and pork juices from collected food waste would be impossible, and so by default they would be eating cannibalistically, which is far from desirable.

The feeding of food waste to pigs – in particular past-their-best vegetables – is a documented risk factor in the susceptibility of pigs to clostridial diseases such as botulism and tetanus, where it's not the organisms *per se* that cause the disease, but the toxins they produce, which upon processing would be released. There is also the issue of mycotoxins from cereals. These toxins are not destroyed through feed manufacturing processes and are chemically stable and both heat and storage resistant. How will these be monitored? Outdoor-reared pigs are more susceptible, particularly to clostridial disease, and as the push is for more pigs to be kept extensively for perceived better welfare of the animals, consideration of feeding vegetable and cereal waste is even more important, especially if they have passed their best and contain both seen and unseen moulds. Some species of moulds have been proven to be capable of causing sudden death.

Diseases eradicated in the UK and other countries also require consideration. Waste food collected from individuals will contain up to 60 % of its meat content from imported meats –meat that potentially contains infectious disease-causing organisms. So how will the inactivation of such organisms in food waste be monitored? What if a batch slipped through with 'cold spots' where the organisms weren't effectively killed? What about new and emerging micro-organisms? It isn't beyond reality that a new heat-resistant organism could emerge, e.g. prion-like organisms – the BSE crisis taught us that.

Even assuming infective organisms were killed 100 % of the time, there could still be implications in feeding them to pigs in the UK. Although a remote possibility, dead organisms or parts of dead organisms could potentially invoke an immune response, particularly those organisms that are naturally transmitted orally or via aerosol. The specific disease-free status of our pigs in the UK means that they are highly sought after worldwide, especially in the breeding programmes of developing and developed nations. Disease-free status is commonly certified by serology surveillance, e.g. after the UK FMD outbreaks. The UK's successful disease eradication programmes, e.g. in respect of Aujeszky's Disease, have used serology to continually prove our disease-free status. It would be a real concern if surveillance showed evidence of a particular disease being present in the UK, caused by the feeding of dead organisms from imported meat from countries where the particular disease of concern is endemic.

One of the most frequently asked questions at a farmers' market is: 'What are your pigs fed on?' How would we go about answering that question honestly with all that has been discussed in this section? Whilst alternative feeding ideas are welcome, they have to be workable, cost effective, thoroughly well thought out, and they should not jeopardise our national herd security. The potential of dangerous ramifications of The Pig Idea cannot be underestimated. The message 'Don't break the swill ban' must remain until these concerns are fully and robustly addressed. Additionally, since the ban on feeding swill has been implemented, the quality of our pork meat has risen dramatically, it's now much easier to regulate the levels of fat, and the meat has a less greasy, cleaner texture.

TIP

Know what you will do instantly during these times so you don't waste your time wondering what to do and then in your haste do something detrimental to the pigs, creating a time-eating, expensive welfare issue to sort out.

NOTE: We are one ham sandwich away from another disease disaster.

CONTINGENCY PLANNING

You will need to have contingency plans in place for various situations, including times when you can't care for your pigs, periods of hot or icy weather, what to do in case of disease outbreaks, unwell pigs, farrowings and disposing of dead stock. You will get tips through the book to assist you in deciding what these will be.

SUMMARY

What has just been described is a basic level of husbandry which is focused on the day-to-day care centered on the pigs' comfort, happiness and health. If you buy a few weaners per year for fattening, and you buy from a breeder who practises a high level of husbandry and health care, then the basic level is more than adequate. You can move to the next level if you wish. The highest standards should be the norm for all pig breeders, although it's not always the case at the moment in the smallholding world, but I think we can learn from the importance the commercial pig world places on biosecurity and disease prevention. They just wouldn't bother with the expense if it didn't make a difference. The higher health status is covered in the herd health chapter. It builds upon the basics using farm health plans (FHPs) with the strict use of biosecurity procedures and preventative medicines such as vaccination.

TIP

It takes a lot more precious time and money to treat pigs with a disease or condition than it does to prevent it.

WHICH BREED SHOULD YOU CHOOSE?

4

Having understood all the implications, and having decided to keep pigs, you have to choose which breed or breeds you wish to start with. Will they be pets, for the table or for breeding? If you are buying for pets, then you may wish to look at the smaller breeds first, such as the Kunekune or the Juliana pig (USA only), as they eat much less and take up less space. If they are for fattening, then your choice may depend upon what you are trying to achieve, e.g. providing meat for yourself, or selling to a butcher or other customers. On the other hand, you may just like the look of a particular breed.

If you are buying for breeding, factors like mothering ability and fecundity (number of piglets born and raised) become more important. When choosing, consider your local climate and the time of year you will be keeping them. Coloured breeds tend to fare better in harsher conditions and are more tolerant to the effects of extreme weather. Think about your meat customers, too. Different breeds have different meat qualities that may or may not suit the end consumer. If the end consumer is you, that's less of a problem, but if you hope to supply a butcher specialising in rare breeds and you have chosen a cross-breed or modern breed, they are unlikely to want them. The reverse can also apply – a butcher expecting a nice lean carcass will be disappointed if you rock up with a traditional breed carrying too much fat.

If you are undecided, lop-eared pig breeds are usually recommended, as some believe they are generally easier to handle – simply because their ears fall over their eyes, limiting their ability to run at full pelt. Prick-eared pig breeders vehemently contest this and many success stories start with a prick-eared breed. I keep both lop- and prick-eared and there isn't much difference in how they behave. I would, however, highly recommend that you avoid anything with wild boar in its parentage the first few times.

ARE THERE SUCH PIGS AS MINIATURE PET PIGS?

Worldwide there are pigs that could realistically come under the 'miniature pig' umbrella. The smallest is the Juliana pig (Figure 4.1), which grows to between one quarter and one third of the weight (50 lb, 3.5 stones or 22 kg) at maturity of the average UK-bred Kunekune pig. The Juliana is a pedigree breed, with its own herd book in the USA, complete with breed standard. The breeding is tightly controlled, producing colourful offspring of reliable mature heights and weights. The only other breed that comes close in size is the Ossabaw Island pig; this is a feral pig originating, unsurprisingly, on Ossabaw Island just off the south coast of Georgia. Here centuries of living on a small island have produced smaller pigs due to insular dwarfism; lack of food on this small island led to natural selection, with pigs adapting to require less food to survive.

Other small breeds, as compared to the majority of farmed pigs, are of similar size to the Kunekune. These include the Yucatan (Mexican hairless pig), the 'Teacup' Pot-bellied pig and the African Pygmy. There are also a few breeds with a hybrid ancestry in the USA and Europe, such as the Minnesota mini-pig – which is of Pot-bellied pig descent – and the German Göttingen mini-pig, which was produced by crossing the Vietnamese Pot-bellied pig, the German Landrace and the Minnesota mini-pig. These pigs were originally bred, and still are, for the biomedical research industry, to provide small pigs suitable for the laboratory environment. A limited number of ex-laboratory pigs can now be found kept as pets in Europe. A related type can be found in Sweden, and to a lesser extent in the UK. Called the Little Swedish pig, these are a hybrid of the German Göttingen mini-pig and the Vietnamese Pot-bellied pig. Most of these breeds have been selectively bred by using smaller and smaller parents, but all of them are genuinely smaller than the main farm breeds of pig we are used to seeing.

So, having established that there are genuine breeds of small pigs, why is there such controversy surrounding the so-called 'micro-pigs' in the UK? The problem is twofold: there are people who want a small pig as a house pet; and there are unscrupulous breeders waiting to dupe them into buying unsuitable animals, and who promote the idea of keeping what is essentially a farm animal inside the house. The buyers tend to have no experience of livestock and wish to have an unusual pet – but one that they can keep as if it were a dog. They have been led to believe they will be buying a pet that

FIGURE 4.1 A Juliana pig.

Photo courtesy of the Juliana Pig Registry [South House Juliana pigs].

will grow no bigger than the average Labrador and which will seamlessly slot into household life. They may well have been tempted by the adorable pictures of piglets sitting in oversized teacups – piglets that are probably only days old, or by compelling media images of Paris Hilton or the cast of *The Only Way Is Essex* taking pet pigs for a walk.

The breeders of so-called 'micro-pigs' in the UK have not imported the Juliana pig or any of the other genuine smaller breeds that exist, but have created their own through runt-to-runt inter-breeding, using breeds that are found in the UK – most often involving the Kunekune. Runts are often the weakest pigs in a litter and should never be bred from – let alone bred to one another. The health of some 'micro-pigs' being sold is a serious concern due to inbreeding depression – a reduction in biological fitness as a result of breeding closely related animals (see Chapter Eight).

There is no established breed of 'micro-pig' in the UK, and so there is no guarantee that the pig purchased will actually stay small. Pigs can reproduce years before they fully mature and unscrupulous breeders parade the parent pigs as fully grown. Some even fraudulently guarantee the piglets will stay under a certain mature weight or height, knowing that by the time the owner realises they own an enormous pig or two, they will have become emotionally attached. Breeders will often sell these herd animals knowing they will be kept alone, which is an infringement of one of the Five Freedoms (see Chapter Three). The worst of the breeders also recommend a diet that starves the animal and unnaturally stunts its growth, and will wiggle out of 'guarantees' if the owner doesn't observe their damaging nutritional guidelines. There have been numerous cases of breeders irresponsibly selling uncastrated males, which can start to display some horrendous, even dangerous, dominant behaviour reasonably early in life. As with all animals that naturally live in groups, there is a pecking order, and should you find yourself lower in the order than the pig, you will seriously have your work cut out in living together harmoniously.

Even as a 'big pig' breeder, I can see how pigs could potentially make great pets; they are intelligent, trainable and interactive, plus they're naturally clean in their toilet habits. However, all pigs – no matter what their size – are livestock; they need company of their own kind and freedom to express natural behaviour. This natural behaviour includes digging and rooting to obtain essential nutrients and to fulfil what is a deep instinctive urge; wallowing to enable temperature regulation and keeping the skin healthy; and full body scratching as a part of their grooming process. None of these behaviours can be expressed whilst being kept in a domestic house. Pigs will turn to tearing up flooring, pulling off skirting boards, ripping radiators from water pipes, destroying fitted kitchens, demolishing sofas . . . the list goes on. As they get older, their strength increases, so what at first is an uber-cute piglet rooting through its pink blanket in its pink dog bed, can, within months, turn into a digging, destructive menace. Pigs are highly intelligent, and highly intelligent animals get bored very quickly. This can lead to delinquent behaviours, such as biting, chewing and barging.

Some owners faced with a pig or pigs that grow bigger than anticipated will try to rehome them. They then find out that this is nigh on impossible and that no one wants them, including the original breeders. So they are faced with either having them shot and incinerated – which, as they have been family pets, is difficult to do – or sending them to a market. The pigs are then sold on for pet-meat production and they would possibly have been better off just being shot and disposed of in the familiarity and comfort of their own surroundings. Farm pigs are used to the hustle and bustle of a farm, they are used to being kept in largish, noisy groups, and so a market is less stressful than for one that has been fed out of a bowl, taken for walks and allowed to sit on the sofa of an evening. It must be terrifying for them. Markets are often the destination for some breeders' piglets that haven't sold. To keep the price artificially high, they will often put their less colourful – and harder to sell – piglets through the markets. Honest breeders just wouldn't breed if there were no customers for them. At least with farm pigs there is always a butcher on stand-by.

If you want a pig as a pet, then that is of course absolutely fine and understandable. Pigs are completely adorable, but choose a breed that reliably grows to a predicted size, such as the Juliana or the Kunekune. Understand that it's going to be a long commitment, as pigs can live for at least 15 years if kept healthy. Buy at least two females or castrated males. Provide them with warm, strong housing with free access to outside space, room for a wallow and shade from the sun. Feed them a proper pig food and not illegal household scraps or pet food – something that I have seen recommended on some American websites. Do your research on how to look after them properly, and know when to deworm and vaccinate, for instance. Also note that you will have to follow the same rules as farmers when keeping pet pigs, including not being able to bury them in the garden when they die. The British Kunekune Pig Society is more than happy to provide expert advice on UK pet pig ownership and even has a rehoming page. In the USA, the Juliana Pig Association and Registry or American Kunekune Pig Society would be good places to start.

Most smallholders and hobby farmers choose a traditional or heritage breed to rear for the table, aiming for a slow-matured tastier meat. It is purported that 95 % of the taste comes from the fat in these traditional breeds and so a modest layer of fat is desirable. Breeds originating from the UK can be found worldwide, but there are also other countries, particularly the USA, that have their own indigenous breeds in need of conservation. Some of the UK traditional breeds have been adapted after being exported and have been prefixed with the country as a part of their breed name, e.g. Berkshire and American Berkshire.

RARE BREED CRITERIA

All of the traditional breeds in the UK are on the Rare Breed Survival Trust (RBST) Watch List and may be marketed, when from registered parents, as pedigree and rare breed. The Livestock Conservancy (TLC) in the USA also maintains a watch list, which includes some of their native breeds, as do other trusts around the world, e.g. the Rare Breeds Trust of Australia (RBTA) and the Rare Breeds Conservation Society of New Zealand (RBCS). The rarity of a breed is defined by the number of registered breeding sows in a country/number of registrations per year and so each breed may move between different watch-list classifications as sow numbers dictate (Table 4.1).

TABLE 4.1 The RBST and TLC criteria for rare breed pigs.

RBST CATEGORY	TOTAL NUMBERS OF UK SOWS	TLC CATEGORY	PIGS
Critical	Less than 100	Critical	Fewer than 200 annual registrations in USA – est. global population less than 2000
Endangered	100–200	Threatened	Fewer than 1000 annual registrations in USA – estimated global population less than 5000
Vulnerable	200–300	Watch	Fewer than 2500 annual registrations in USA – estimated global population less than 10,000
At risk	300–500	Recovering	Breeds that were once listed in another category but are still in need of monitoring
Minority	500–1000	Study	Breeds that are of genetic interest but either lack definition/genetic or historical documentation
Breed density and distribution can mean a breed will be highly vulnerable to disease outbreaks, and are included in making assessments of endangerment		Also included are breeds that present genetic or numerical concerns or have a limited geographic distribution	

TRADITIONAL AND HERITAGE BREEDS

Berkshire

Rare breed classification: RBST: Vulnerable

Meat type: Mostly known for their succulent pork and crunchy crackling in the UK

FIGURE 4.2 A Berkshire pig.

Photo courtesy of Practical Pigs magazine.

This breed was once known as the 'lady's pig' because of its easy-to-handle nature. They are prick eared and colours range from very dark brown to black. They have six white points, one on the head, one on each foot and one on the tip of the tail. They are lively pigs, bordering on cheeky when young, but not aggressive to handle. They have good-sized litters of chunky-looking piglets and are careful mothers. They love being kept outdoors and they are one of the hardiest breeds should you live somewhere a bit more rugged. They make absolutely delicious pork; the Japanese pay a premium for Berkshire 'Kurobuta Black Pork', which, contrary to the name suggests, doesn't look black when processed. Berkshire pork has been scientifically proven to have better colour, texture, marbling, ultimate pH and water-holding capacity – all known factors to better eating quality and making pork the meat of choice. In the USA, Berkshire pork can attract a substantial premium from pork dealers. Berkshires often reach pork weight a tad earlier than other breeds and, due to their fine bone, produce a high proportion of their weight as meat. Another bonus with this breed is that they can be fed slightly less than other traditional breeds and still maintain their weight, but they're inclined to run to fat when older, so do not lend themselves to bacon so well.

British Lop

Rare breed classification:
RBST: Vulnerable

Meat type:
Dual-purpose pork and bacon

FIGURE 4.3 A British Lop pig.

Photo courtesy of Practical Pigs magazine.

This breed, at present, is one of the rarest pedigree pig breeds in the UK. In fact, there are three times as many Giant Pandas in the world! They are all-white pigs with, unsurprisingly, lop ears. They have a similar appearance to the Welsh and British Landrace pigs and indeed these pigs are in their ancestry, but they are genetically distinct and visually distinguishable from each other, once familiar with the breeds. They're an obedient, resourceful pig, happy being kept inside or outside. They can tolerate extremes of temperature but their white skin needs ample protection from the sun. They are excellent mothers, producing lots of milk for their large litters. They are economical converters of feed for both pork and bacon and do not run to fat easily. The lop hasn't been a favourite among smallholders due in most part to its plain white appearance, but perhaps it should be, as it would suit the rare breed niche market and commercial butcher alike.

British Saddleback

This breed is also colloquially known as the 'gentleman's pig' and the 'farmer's pig', and can be instantly recognised by their semi-lop ears and a white band of hair over the shoulder, extending down both front legs. They may or may not have some white on their back legs and on the tip of the tail. In Australia there is a related subset of the breed known as the Wessex Saddleback. Although originally an imported breed from the UK, the Wessex has been shown to be genetically distinct. The Saddleback is widely acknowledged as the best 'dam' (mother) pig, with an excellent mothering ability, large litters

Rare breed
classification:
RBST: At Risk, TLC:
Study, RBTA: Critical

Meat type:
Dual-purpose pork
and bacon

FIGURE 4.4 A British Saddleback pig.

Photo courtesy of Practical Pigs magazine.

and lots of high-quality milk to produce robust, healthy piglets. They are hardy pigs and can thrive where other breeds would not and are also very docile to handle. The meat is dual purpose and can produce both succulent pork and decent bacon if not over-fed.

Oxford, Sandy and Black

Rare breed
classification:
RBST: At Risk

Meat type:
Dual-purpose pork
and sweet bacon

FIGURE 4.5 An Oxford, Sandy and Black sow.

Photo courtesy of Barnsnap pigs.

These pigs are fondly known as 'plum pudding' pigs due to their striking appearance. This is a revived breed in the UK with a pedigree herd book dating from the mid-1980s.

The pigs are a semi-lop-eared breed and their colour ranges from a sandy brown to a vibrant rust colour with black patches (not spots or flecks) with a white blaze on the face, white tail tip, and pale/white coloured feet. Evidence of a similar-looking pig dates back a few hundred years, but as no pedigree had been officially recorded, the OSB could not, until fairly recently, be regarded as rare breed or pedigree. Thanks to dedicated breeders, the breed is now formally accepted as pedigree and has been recently added to the RBST watch list. The OSB is generally a docile, friendly, laid-back breed. They make good, attentive mothers with decent milk. Like the other darker pigs, they are more resistant to sunburn, so their summer care is somewhat easier. They lay down less fat than some other breeds so are excellent for bacon as well as sweet lean pork.

Large Black

Rare breed classification: RBST: Vulnerable, TLC: Critical, RBTA: Critical

Meat type: Dual purpose but lends itself to bacon better than some other breeds

FIGURE 4.6 A Large Black pig.

Photo courtesy of Practical Pigs magazine.

These are sometimes called 'Cornish Blacks', in reference to one of their ancestral origins, or 'elephant pigs' due to the look of the newborn piglets with their long lopped ears. They are the UK's only all-black pig and they have been exported worldwide, including to the USA and Australia. In the USA, they have a large following of supporters. They are docile to handle, contented with life in general and extremely good mothers. They are a little more economical to rear if you can provide a reasonable-sized pasture for them to graze on and, as they have fewer tendencies to dig than other breeds, they may not even destroy it. They can tolerate the sun more than other breeds as they don't have a speck of pink skin to burn. There are conflicting opinions as to whether they run to fat easily or not, but there is no doubt they make a tasty pork and even better bacon. I personally really like this breed and always admire them when I see them in the show ring.

Gloucestershire Old Spots

Rare breed classification: RBST: Minority, TLC: Critical

Meat type: Dual-purpose pork and bacon

FIGURE 4.7 A Gloucestershire Old Spots pig.

Photo courtesy of Practical Pigs magazine.

The eminently named 'orchard pig' is probably the best known of the breeds. A predominantly white breed, they must have at least one clearly defined spot that doesn't merge with other spots. Like other traditional breeds, they are docile and, despite their white skin, they are hardy when provided with a suitable environment. They are good mothers, which produce large litters of strong piglets and have lots of good-quality milk. As they get older, we found their very lop ears makes them almost blind, which in turn can make them exasperating when you are trying to make them do something even moderately quickly or when they are mixed with other breeds. From a pressured time point of view, despite their undeniable quality, they may not be the best breed to choose. Like the British Saddleback, they produce excellent dual-purpose meat that lends itself equally as well to pork and bacon when carefully fed.

Middle White

These beautiful completely white pigs are also referred to as the 'London Porker' and the 'ugly breed'. In the USA, Canada and Australia, they are called the 'Middle Yorkshire'. They are compact in stature with an exquisite turned-up snout. They need protection from the extremes of heat and cold weather if they are kept outdoors and, due to their short legs, do not enjoy being kept in thick mud in the winter. These very sweet-natured, friendly, sociable, talkative pigs are one of the most passive of the prick-eared breeds to handle. They are best kept for their excellent pork rather than for bacon and I would place a bet with anyone that no other breed can beat the crackling on the Sunday joint.

FIGURE 4.8 Our son Oliver showing his Middle White pig.

The Japanese crave the pork enough to have erected a statue in its honour and the meat has won blind taste tests on multiple occasions. Like the Berkshire, the Middle White are slightly finer boned and can be sent to the abattoir slightly earlier than other breeds, with a high percentage coming back as meat, even as high as 90 % has been reported. I have heard that they can be a little more prone to squashing their piglets but I have zero evidence of this and our small herd has not (yet!) squashed any more than our British Saddlebacks.

Tamworth

FIGURE 4.9 A Tamworth pig.

Photo courtesy of Practical Pigs magazine.

The 'forest pig' or the aristocratic pig for all seasons, the Tamworth is worthy of a place on a smallholding, even if only for its efficiency in turning over a plot of land ready for planting veg. Tamworth is a prick-eared breed, ginger in colour with an almost wild-boar-length snout. Genetically, they are the purest of breeds with the greatest genetic diversity in the UK, so it's no surprise that they thrive when kept outdoors in any weather. If you are thinking of rearing pigs indoors for long periods of time, don't choose this breed as their inquisitive characters and high intelligence will make this type of husbandry too confining. As with all pigs, they are not vicious to handle although they can be very lively. In fact, their forward-going spirit might even be an asset to those with time pressures. They are a dual-purpose pork and bacon pig and have won a scientifically controlled taste test as the best-tasting pork. It should be noted here that the 'Tamworth Two' that infamously caused mayhem after escaping on their way to an abattoir in the UK, and earned the Tamworth an unfair reputation, were actually 50 % wild boar, and in my opinion probably weren't socialised very much by their keepers.

HERITAGE BREEDS INDIGENOUS TO THE USA

Ossabaw Island Pig

Rare breed classification:
TLC: Critical

Meat type: Gamey pork

FIGURE 4.10 An Ossabaw Island pig.

Photo courtesy of The Livestock Conservancy.

There are a few breeds that are unique to the USA and are rarely, if ever, found overseas. Some are so rare they were only found in strict geographical areas, such as the Ossabaw Island pigs, which were until 1970s exclusively found on an uninhabited island off the coast of Georgia, with exportation to the mainland forbidden. By all accounts, you should get to the front of the queue to own one, as the meat is reported to be a beautifully marbled dark red, with a rich, wild flavour that commands the attention of gourmet chefs.

Choctaw

Rare breed classification:
TLC: Critical

Meat type:
Full-of-flavour pork and charcuterie

FIGURE 4.11 A Choctaw pig.

Photo courtesy of The Livestock Conservancy.

These pigs are of Spanish origin but are now an American breed mostly confined to the south-east Oklahoma counties. The rarest of the rare breeds in the USA, these were once a very important food resource for Native Americans of the area as well as farmers. They are a smaller breed than typical farmed pigs, with pricked or slightly lop ears. They are usually black but may have some white markings; with well-developed shoulders, long legs and narrow hams, they are streamlined for speed and have unrivalled agility. Many have wattles (aka tassels, piri-piri) and they have a single fused hoof known as a mule foot (similar to a horse's hoof). Typically they're kept as free-range pigs; however, their rare breed status has seen a slow increase in the numbers kept by rare breed enthusiasts, where they readily adapt to being domesticated. There is no formal breed registry but the Livestock Conservancy monitors their numbers. Their small size means that if meat production is your primary goal, then these may not be an appropriate choice. The meat they produce is dark and flavourful, often used for the production of sugared hams.

Mulefoot

Rare breed
classification:
TLC: Critical

Meat type:
Gourmet pork

FIGURE 4.12 A Mulefoot pig.

Photo courtesy of The Livestock Conservancy.

The Mulefoot origins are thought to be Spanish, but as they have been inhabitants of the USA long before herd registers and pedigree recording took place, no one actually knows. They are named after the structure of their hoof, which is single, fused and non-cloven. They are compact in appearance, solid black in colour but may have some white in their coat and they have semi-lop ears. Occasionally they have piglets with wattles under the chin, indicating a common ancestor to other wattled breeds or mixed heritage early in their development. The breed is hardy and resilient, calmly producing between 5 and 12 piglets per litter, and has a gentle nature when handled. The meat is considered gourmet 'moist and melt in the mouth' with a particular leaning to making a high-quality ham, and so may provide a useful income if there is access to local, high-end restaurants looking for a specialist niche product.

Guinea Hog

These black pigs have a variety of regional names including the Acorn Eater, Yard pig, Pineywoods Guinea and Guinea Forest Hog. Recent DNA-based evidence supports the theory that the Guinea Hog derived from the old Essex line of the British Saddleback breed of pig and was once known as the 'Improved Essex' or 'Fisher Hobbs'. Their snouts can vary from long to short but most have black to bluish-black hairy coats, pricked ears and curly tails. They are a highly resourceful, robust, fearless breed. Given a large area to roam, they can produce meat and lard with minimal human input. When reared on

Rare breed
classification:
TLC: Critical

Meat type: Traditional
pork and charcuterie

FIGURE 4.13 A Guinea Hog.

Photo courtesy of The Livestock Conservancy.

the farm, they are friendly pigs to handle and will root and dig tirelessly and will even rid the farm of snakes and rodents without losing pace. The meat has a traditional, rich taste and makes the most fabulous charcuterie products. The pigs can fatten very easily when fed too much of a domestic diet and careful monitoring of intake is required.

Red Wattle

Variants of the Red Wattle, with individual prefixes added to the name, have been bred over the years across the USA, but this has been done carefully to retain the classic appearance of the pig. They are of a similar size to the major farm breeds and come in a variety of red colours from a pale red right through to almost black and they may or may not have black speckles or patches in their coat. They are stocky in appearance with slim heads, long snouts and a slightly arched back. Under their jowls they also have two wattles hanging down giving them their name. Red Wattles have an easy-going personality with a good fecundity rate of 10 to 15 piglets per litter. They are a hardy, coloured breed, excellent foragers given the space to do so, and have a rapid growth rate. The meat is distinctive and meaty in taste, making this a definite contender for the small-scale keeper.

Rare breed
classification:
TLC: Critical

Meat type: Meaty pork

FIGURE 4.14 A Red Wattle.

Photo courtesy of The Livestock Conservancy.

Hereford

Rare breed
classification:
TLC: Watch

Meat type: Lean but full
of flavour

FIGURE 4.15 A Hereford pig.

Photo courtesy of Prairie State Semen, Inc. and www.showpigs.com.

A striking, much admired large breed of pig, they are predominantly red-brown colour with a white face, feet and sometimes undercarriage and specifically bred to resemble Hereford cattle in the UK, with similar colours and markings. They are purported to be 'the most beautiful pigs in the world'. They have semi-lop ears, a wide, slightly dished face with a slight arch to their back. They are hardy when given shade to protect their white markings and suited for either extensive or intensive production systems, with a docile, compliant temperament, making it a favourite breed to use in youth programmes, e.g. 4-H, FFA. Their mothering skills mean they rarely squash their piglets and they reliably have eight to nine piglets per litter. If you decide that they are worth keeping, the meat is easy to keep lean with moderately careful feeding without any loss of texture and flavour – although, as with any animal, excess feeding will cause additional fat to be laid down.

MODERN BREEDS

The definition of a pig as modern seems to be in its suitability for, and/or genetic adaption to suit the larger-scale, intensive commercial systems. Large healthy litters and fast growth rates are the primary goals. Three of the UK modern breeds, the Large White, the British Landrace and the Welsh, are also officially rare breeds in the UK and are worthy of preserving. Most modern breeds can be found across the world, sometimes with the country of residence as a prefix to the breed name, e.g. Duroc, American Duroc. Differences in breeding programmes will have made them diverge genetically to a greater or lesser extent over the years, but their desirable, inherent characteristics will be similar.

Large White

This breed is commonly known around the world as the 'Large Yorkshire' pig, in reverence to its ancestral origins in the North of England, and it is also a foundation breed of

Rare breed classification: RBST: At Risk

Meat type: Dual-purpose lean pork and bacon

FIGURE 4.16 A Large White/Yorkshire pig.

Photo courtesy of Prairie State Semen, Inc and www.showpigs.com

some other breeds we still see today, e.g. the Middle White. It is called a modern breed due to its development for modern use, and the subsequent influence and ability to stamp quality into modern breeding programmes.

They are most often used as foundation breeding stock for producing high-vigour hybrid pigs, as they consistently have large litters of fast-growing piglets and produce lots of milk or sire numerous fast-growing piglets. Large Whites are completely white but easily distinguishable from other white breeds due to their large prick ears and long, slightly dished snout and they are an easy pig to manage. Even though they have a fine hair, they can be kept outdoors successfully if strictly protected from the extremes of the weather, but are better suited to warmer indoor systems. Pedigree Large Whites may be harder to source than some other rare breeds, as they are mostly used for the production of hybrid pigs, but their use in their current form is getting rarer as high-tech hybrids are further developed. Dedicated breeders are trying to remedy this, so they are not pushed further into the rare breed category. The meat is lean and the pigs do not pile on fat when fed correctly, making them popular with the typical butcher. Currently, they are not a favourite within the smallholding fraternity as they are associated with supermarket meat, which is exactly what they are trying to avoid. Those who own them, however, say they are the best pig they have ever kept and wish they had kept them sooner, so perhaps give them a try!

Welsh

The Welsh pig was once one of the most popular breeds of pig in the UK, being hardier than the Large White and able to thrive outdoors without losing its fast growth. They are completely white in colour with large lop ears, long in length with massive hams (rear

Rare breed classification: RBST: At Risk

Meat type: Dual-purpose lean pork and bacon

FIGURE 4.17 A Welsh pig.

Photo courtesy of Practical Pigs magazine.

ends). They are particularly good mothers and rear their litters well. They have amiable characters and are gentle to handle and very well suited to the smallholder wishing to rear a leaner meat than a traditional breed in outdoor conditions. They are said to be of similar quality to the British Landrace and British Lop in terms of high-quality lean meat and can used to produce both pork and bacon.

British Landrace

Rare breed classification: RBST: Endangered

Meat type: Dual-purpose lean pork and bacon

FIGURE 4.18 A British Landrace pig.

Photo courtesy of Practical Pigs magazine.

This breed was originally bred in Denmark for use in commercial bacon production. It is a white pig with semi-lop ears, a long snout, long lean body with an extra rib, and fine hair. They are fast growing and capable of producing a very lean pork at an earlier age than most traditional breeds. They are easily managed in an indoor system with less human interaction, and produce large litters and high-quality milk. They are rarely kept outdoors and so perhaps are less robust than other breeds in varying climates. I would be cautious of keeping them outside due to their fine coats, and also because of the leanness of the meat they produce, but I have never owned one or tasted one reared the smallholder way, so I don't have any personal knowledge and may be unduly cautious.

Piétrain

These muscular, chunky pigs originate from Belgium, where in a hybrid breeding pro-gramme a genetic mutation was identified and commercially exploited. They have semi-lop ears and are predominantly white with black spots with a wide set square

Meat type: Dual-purpose lean pork and bacon

FIGURE 4.19 A Piétrain pig.

Photo courtesy of Mark Horsley.

appearance in the body. They have a high lean-meat-to-bone ratio, with enormous double-muscled hams and are typically used to improve the lean meat content of other commercial breeds. You have to be careful with this breed, as some of the bloodlines can drop dead if stressed, due to a faulty gene, meaning these are neither hardy nor vigorous. There has been considerable investment by the industry to remove this 'halothane' or 'stress' gene (Chapter Eight) from the breed and there are now bloodlines free of the problem. Although the Piétrains I have met have been friendly, affable and easy to handle, because of their extraordinarily large hams and short legs they may not naturally give birth easily and are probably best left to the professional breeder. Why give yourself the headache of potentially increased birthing complications? If you can find a breeder within a reasonable distance, the meat is very lean and tasty.

Duroc

This breed originated in the USA and can be kept outdoors with great success. They have a high meat-to-feed ratio and are often used in commercial systems to add 'meatiness' to the flavour of other breeds. They have a beautiful deep auburn colour through the body and shine like a conker when well kept. They have smaller heads than some other breeds and erect ears that fold over at the tip. They are reported to be an excellent smallholder fattening pig by those who breed them but much misunderstood by those that don't; they can be affectionate and interactive. They are generally calm for routine handling and easily contained by non-electric fencing. They are quick growing and do not lay down much fat, as you would expect from a modern breed. The meat is full of flavour and the little fat it does have is marbled inside the meat, which bastes the joint from within when cooked. Before you get carried away with this idyllic-sounding

Meat type:
Dual-purpose lean
pork and bacon

FIGURE 4.20 A Duroc pig.

Photo courtesy of Practical Pigs magazine.

breed and order yourself half a dozen, they are fearsome protective mothers and there is absolutely no option to be able to 'assist' during the birth or even touch the piglets with the sow present, e.g. if a wound needed seeing to. So while rearing fattening weaners wouldn't be a problem, perhaps the breeding of these pigs should be given further thought if you're time restricted.

Hampshire

These are known as the 'thin rind' pig. They are an active breed, producing maximum meat with little fat when fed correctly. To the novice, they look like British Saddleback

Meat type:
Predominantly bacon

FIGURE 4.21 A Hampshire pig.

Photo courtesy of Practical Pigs magazine.

pigs with prick ears, but in truth the only similarity is the colour markings. They are heavily muscled, long-legged pigs originally developed in the USA for use in commercial breeding programmes to improve growth rates and meat-to-fat ratios. Although not normally eaten as a pure breed, the meat is very lean and makes good cured bacon. Before deciding they are the pig for you, be aware that working boars can be aggressive to handle, which could eat up some of your valuable time when handling and even in performing husbandry duties. Despite this, they are slowly increasing in popularity as a bacon pig among the smallholding community and do well kept outdoors.

MODERN BREEDS INDIGENOUS TO THE USA

Poland China

The Poland China is one of the oldest American breeds, originally developed in the 1800s for its size and ability to travel long distances to market. Today they are recognised as the leaders in American pork production. They are a lop-eared, black-coloured breed with six white points on their face, feet and tail. They possess a large, thick-boned but long frame, which is both muscular and lean, lending itself well to commercial pig production. The Poland China is an excellent converter of feed, gains weight readily with good husbandry, and can be kept in any system with ease. It is quiet in temperament but robust and rugged in its constitution.

FIGURE 4.22 A Poland China pig.

Photo courtesy of Prairie State Semen, Inc. and www.showpigs.com.

Chester White

Chester Whites are classified by their followers as heritage hogs and are known for superior mothering abilities, durability and soundness. They are completely white, with a dished face, medium fall (semi-lop) ears with a full thick coat. Chester White females have long been known for a high conception rate and a large number of vigorous pigs per litter. They are known for remaining productive for more parities than other modern breeds. The Chester White is a versatile breed, suited to both intensive and extensive husbandry systems, although they are prone to sunburn and access to continuous shade is essential.

Chester Whites have been preferred by packing houses for their superior muscle and white skin, which dresses out to a light pink. Their feed conversion rates are high, gaining up to 1.36 lbs (0.62 kg)/day and produce 1 lb (0.45 kg) of lean meat for every 3 lbs (1.4 kg) of feed, and so they will remain an important breed in the USA.

FIGURE 4.23 A Chester White.

Photo courtesy of Prairie State Semen, Inc. and www.showpigs.c

Spotted Swine

These pigs are fondly known as 'Spots' in the USA and at least part of their ancestry can be traced to the original Poland China pigs, with some Gloucestershire Old Spots heritage. They are an attractive black and white colour with lopped ears and strong stocky bodies.

FIGURE 4.24 A Spotted Swine.

Photo courtesy of Prairie State Semen, Inc. and www.showpigs.com.

Spots have continued to be improved in terms of feed efficiency, rate of gain and carcass quality, as can be proven in the testing stations throughout the country. Spots are popular with farmers and commercial swine producers for their ability to transmit their fast-gaining, feed-efficient meat qualities to their offspring. The Spots female is highly productive, known to raise what she farrows while keeping her condition and without sacrificing her gentle temperament. It is thought by those in the know to be under-rated as a commercial pig. The boars are keen to work and 'don't need a diagram' to get it right from early in their career. These rugged pigs are suitable in the weather-extreme USA states and are worthy of consideration.

OTHER PEDIGREE BREEDS

Mangalitza

These woolly pigs are sometimes referred to as the 'sheep', 'lard', or 'charcuterie' pig and originate from Hungary. They can be found across the globe and come in three distinct colour types – red, blond and swallow bellied (black with a pale blond undercarriage) and all have a dense curled hair that resembles wool. They are a late-maturing but hardy, disease-resistant breed and live outdoors with ease.

They usually have smaller litters and are extremely protective mothers who will attack at the slightest provocation when they have piglets. The red Mangalitza is said to be slightly more docile than the blond and swallow-bellied varieties. In my limited

FIGURE 4.25 A Mangalitza pig.

Photo courtesy of Practical Pigs magazine.

experience of the breed they all have a tendency to nip when investigating and bite when challenged. They are endearingly stripey when young, and will find any chink in your fencing and explore other areas of your farm. Their fat can be whipped into lard and they are primarily used in the production of processed charcuterie, e.g. salami and chorizo. They can live together all year round, with the boar, and form close family groups, rearing one another's piglets without aggression to each other. They should be fed slightly lower quantities than quoted in this book as they mature very slowly and could run to excessive fat. Some breeders claim that they are Lincolnshire Curly Coats (an extinct breed) and say that you will be helping to revive the breed if you breed them. They are not, however, the only similarity being that both breeds share the genetics that create a curly coat. Mangalitzas are reasonably uncommon in the UK and USA but they are bred in greater quantities in other parts of Europe.

Kunekune

A small breed traditionally kept as pets in the UK, USA and around the globe. They come in a range of colours and may or may not have piri-piri (aka wattles, tassels) under their chins. As they are smaller, they eat correspondingly much less than the quantities quoted in *Commuter Pig Keeper* and so are cheaper to keep; they are also good grazers with fewer tendencies to root and dig. They are late maturing and produce quite small litters, but contrary to popular belief they were originally a fattening pig and can be used for meat. It is a slightly fattier meat than the traditional breeds but the sausages and charcuterie they produce benefit from this. The commercial world has started to take notice of this breed, even using the boars to cross with commercial white pigs, improving the flavour of the meat. Like the Mangalitza, they can be used for the production of lard,

FIGURE 4.26 A Kunekune pig.

Photo courtesy of Practical Pigs magazine.

but they can also make excellent sausages and tasty pork joints. Remember, however, that the carcass takes longer to reach pork weights, so you will need to plan this into any meat production plans.

OTHER NAMED BREEDS

Although not recognised breeds as such, you may see Iron Age (Figure 4.27) and wild boar advertised for sale. People sometimes use the terms synonymously, but they are very different. Wild boar are not suited to smallholders, not least because in the UK you need a Dangerous Wild Animals Licence to keep them, while in the USA, feral pig keeping is subject to additional state restrictions.

They also have temperaments you would not want to mess with. Iron Age pigs are the result of crossing wild boar with the Tamworth and they are now visibly breeding true. Their temperaments can be similar to that of wild boar, in that they are unpredictable, with a tendency to bite, and are extremely protective mothers of their stripey offspring. They seem very similar to the Mangalitza in their personalities but perhaps a little less easy to manage. I have tried the meat and did not find it remarkable at all, even bland if truth be told. Personally, I would avoid them as they are not pedigree, so you really don't know what you are buying or breeding from.

You will also see Vietnamese Pot-bellied, Swedish Pot-bellied, and Meishan pigs for sale, and although not recognised as pedigree pigs in the UK, they often are in their countries of origin. If the UK breeder has bred them true and not crossed them with another breed, you will have a good idea of their final size and suitability to your venture.

FIGURE 4.27
An Iron Age pig.

*Photo courtesy of
Liz Shankland.*

SELECTING AND SOURCING BREEDS

How Much Will I Have to Pay?

The price of registered breeding stock varies wildly between the different breeds and both across and within regions. You can also pay more for a particularly good example of the breed, for older animals ready to begin breeding, and for 'proven' animals – ones which have produced offspring. As expected, a premium may be charged for very rare bloodlines or a show winner. Buy the best stock you can afford, especially the breeding boar, and have a go at breeding your own show winner – but only after you have mastered the art of rearing pigs and are sure you really want to make a much more considerable commitment. Choose a healthy, strong pig with a good mid-length for carrying those precious piglets. Most breed websites/social media sites give a good idea of the current prices being charged in the different areas.

TIP

If you are breeding, buy the best stock you can afford, particularly the boars, as they provide 50% of your herd genetics.

Currently (2016), for traditional breeds in the UK fattening weaners are around £45 to £65; registered gilts at 8 to 12 weeks are around £70 to £150; maiden gilts at 6 to 12 months, £160 to £300; in-pig gilts £200 to £450, and boars ready to work at 6 to 12 months, £180 to £350.

In the USA (2016), fattening (feeder) weaners, non-heritage gilts or barrows, $50 to $75; heritage breeds, $100 to $200; registered gilts, $200 to $400; maiden gilts, $350 to $450; in-pig gilts/sows, $350 to $1500, and boars ready to work, $250 to $500.

You will see stock at various ages advertised much cheaper than the average in an area; it may be a genuine herd reduction and the need for the pigs to go is greater than the need for money; the owners could be eccentric millionaires; but they also could be cutting huge corners on health and welfare. Look around all the livestock on the farm, ask about their deworming and vaccination programme, ask to see the medicine book and make your own mind up about the health status of all the pigs – not just the ones you are considering buying. If they say they have been dewormed or vaccinated and you don't wish to look untrusting, just ask for the exact dates they were done and they should immediately look it up in their medicine book.

Selecting a Breed for Taste

I think it is an excellent idea to buy different breeds of pigs each time and use the experience to find out which you like the taste of and that you like to keep. We didn't do that and I now wish we had, and although we love British Saddleback and Middle White pigs, for many reasons I feel we have missed out. There is, of course, nothing stopping us doing it now, except we tend to have enough meat coming through for our needs, and when I go to other breeders' farms, their breed of pork is usually served if I hang around long enough to be fed.

There is nothing stopping you buying different breeds of piglets at the same time, as at 8 to 10 weeks they generally mix well, especially if you can pen them next to each other for a while before mixing together. They will then only have a short scrap and rarely cause each other much harm apart from a few treatable scratches. It does, of course, double the amount of accommodation you will require, but mixing recently weaned piglets straight away in their new home will add considerably to their stress and may bring to the fore an underlying problem that their immune system was previously dealing with. Do the actual mixing just before bedtime with copious amounts of feed available and two troughs of water apart from each other. That way they will go to sleep and wake up next to each other, minimising the scrapping.

I, personally, do not like the typical meat available in the supermarkets, as it just tastes bland and tough to me. It must be a breed or husbandry difference, as we feed the same type of feed. The taste you get out of your chosen breed can be reflective of how you have fed it. This was apparent when I tried some British Saddleback sausages at a local sausage competition, which were just awful. We later found out that the founding stock was originally bred by us so they even had the same bloodline as us and we used the same butcher to make them. Had I known whose sausages they were, I would not have tried them, as I knew they fed their pigs a diet of mouldy vegetables and fruit, old bread, cake, milk and very little purchased pig food. It was definitely reflected in the taste.

Don't dismiss trying cross-breeds. Certain crosses do go well together, e.g. British Saddleback crossed with Gloucestershire Old Spots is known for its tasty pork and crispy crackling. There are as many crosses as you can think of, and it would be quite hard to try them all, but if they interest you, have a go, remembering that a lot of the taste is in how they are looked after and fed.

NOTE: Modern breeds reach pork weight sooner than other traditional breeds. Some traditional breeds, e.g. Berkshire and Middle White, are sent to the abattoir at lighter weights to maximise the percentage of meat to bone. Some breeds take longer to mature, e.g. Mangalitza and Kunekune, but can be fed slightly less than quoted in Commuter Pig Keeper.

Sourcing Pigs

So you have chosen your breed or cross-breed. Now you have to source them. With cross-breeds, you will probably just come across them being advertised. With pedigree breeds, take a look at the breed society websites (find contact details via the British Pig Association or the Livestock Conservancy websites); ask your local farming/smallholding group or pig-course provider, if you attended one, to recommend people in your area; search online advertising sites, smallholder website forums and feed merchant boards; use social media sites; check the 'for sale' pages and breeders' directories in specialist smallholding and pig magazines. Good recommendations often come from word of mouth, and if the breeder you have found doesn't have any weaners at the time you want them, they will usually, if they are decent breeders, recommend someone who has, or you can book from future litters and wait.

I would also be reluctant to buy from someone who tries to 'hard sell' their piglets. Never buy under pressure. You won't enjoy them anywhere near as much. In the back of my mind would be, 'Why are you so desperate to sell these? What's wrong with them?' It would ruin the experience even now, many years down the pig-breeding road.

They may have been underfed, so they have been cold and grown more hair to compensate, which is not a good start; or they may not be converting their feed efficiently, due to a large worm burden. It doesn't matter what the reason is, as long as you don't buy the animal. You could be getting into a heap of trouble. Also don't necessarily go for the biggest one, especially if they have not been weaned from the mother for at least two weeks, as the biggest one could be the most susceptible to 'staggers' caused by a bowel oedema (swelling), sometimes attributed to over-eating. This is covered in Chapter Nine in greater detail.

NOTE: Don't buy from livestock auctions or markets, at least until you know what you are looking for in a pig and you know enough about pig husbandry to be able to treat them for all the unseen problems they may come back with. Yes, you may get them at a knock-down bargain price, but you will have absolutely no idea of their health status.

Buying purebred fattening weaners from pedigree breeders is usually not any more expensive than cross-breeds; they may not meet the breed standard and so cannot be registered, shown, and should not be bred from, but they will still be from pedigree parents. These 'mis-marks', as we call them, might have been ruled out for pedigree registrations because the nipples are out of alignment, they may have prick ears when they should be lop, or they may have incorrect body markings.

An additional advantage to buying from a pedigree herd in the UK is that you can have a pedigree meat certificate for each piglet proving its parentage, so you know exactly what you are eating and selling. You can, of course, buy registered breeding weaners and eat them, but you will pay a lot more for registered stock and you could be eating the potential 'Pig of the Year' competition winner. Some breeders will take orders months in advance and may require a deposit, while others advertise only when the weaners are ready to go. Always check they have been dewormed and with what product before you take them home.

Should You Buy Males or Females?

As fattening weaners are generally in the freezer before the time they are sexually mature, there is little difference in the handling of either sex up until this time. Some people mention boar taint – an unpleasant smell or taste – in the meat of uncastrated boars, but we have never experienced this in ours. There is some doubt that it is as much of a problem in the slow-maturing traditional breeds, although I'm still on the fence on that one as it could just be a very rare occurrence. This is the reason why most pigs raised for bacon by smallholders are female. From all my years of keeping pigs, I have only heard of one case in an older Gloucestershire Old Spots boar kept in France, so it must be very rare. Males often reach pork weight a little sooner, so could be considered a bargain. Castrated males are more commonly available in the USA and are known as 'barrows'. Castration is the UK has to have special veterinary permission to be routinely performed by the farmer.

If you have any doubts about taking your first set of pigs to slaughter, we always recommend that you choose male pigs, as you are more likely to book them in to the slaughter house, instead of being tempted to keep them. Animals that are cute now will turn into huge boars that you cannot do anything with, forcing you to complete what you set out to achieve.

Same-sex piglets of the same breed are likely to grow at a similar rate and will therefore be ready for slaughter at the same time. If you only have two and one is ready for slaughter earlier than the other, then you have a dilemma, as you should not keep a pig on its own. Another concern is if something happens and you have to keep the pigs for slightly longer than you intended, then there will be 'doctor and nurse' games being performed in the pen and the female may be harassed by the male pigs' boisterous behaviour.

Selecting for Breeding

Buying breeding stock is a whole different process. You may select a breed for taste, because you will get quite a few piglets out of her, because you like the look of the breed, because you like the temperament or because you know there is a market for the breed. You can, of course, breed from unregistered pigs. So your favourite 'Miss Piggy', who you really should have sent to slaughter but got too attached to, can technically be bred from in most cases. However, you will be limited to selling the offspring as non-pedigree fattening weaners. Even if Miss Piggy is a pure-bred pig, the breeder will have decided she wasn't good enough to breed from and so she is unregistered. So if you cross Miss Piggy with a registered boar you still cannot register any of her progeny, and would have to market your weaners as cross-breeds, even if your sow *looks* the same breed as the boar.

To give her the best chance of being able to rear piglets, you need to check that she has a reasonably correct nipple alignment for feeding them; if they are all up one end, breeding from her would not be a good idea. It would be advantageous if she has a long, firm, deep body, so the piglets have more room when growing – and, of course, do not breed from her if she has any deformity or obvious weakness. If you wish to breed, pay the bit extra for registered gilts or sows. Even if you don't intend to register any offspring now, you can then change your mind at a later date. Miss Piggy can always be there for company, giving you the excuse to keep her.

When buying registered stock, the weaner, gilt or sow will already conform to the minimum requirement of the breed standard. It is worth having a quick check yourself, especially if you have more than one choice in the litter. Getting to know your breed standard will help you pick the best one. There are, sadly, a few breeders of pedigree pigs who either don't know the requirements or don't care, and who register pigs that do not conform to the minimum standards, so do swot up and even take a copy of the breed standards with you. It certainly wouldn't offend me if you turned up at mine clutching a copy! We have driven miles to buy a pig, only to take one look and say, 'No thanks – she is not breed standard.' We always show our pigs to prospective customers without obligation and do not take offence if they are not what they are looking for. They might like certain types or be looking for a boar to complement the shape of their sows.

Before we became a predominantly closed herd, with all our female breeding pigs now from our own stock, we liked to buy our breeding stock as weaners (8 to 12 weeks old). This way we could manage their feeding and care from a very young age, and it also helped them get to know and trust us, which in turn really helped us when they had just farrowed. We accepted that it would take longer to establish our full breeding herd and we wanted more out of the experience than a high turnaround on the money invested. We have purchased older stock from trusted sources when we wanted certain bloodlines, but the odd one just didn't bond with us in the same way and we were pleased that our original decision to buy them young was proved right . . . for us. We did take a gamble, though, as they could have been sub or infertile.

Registered pigs sell for more than unregistered, so your initial outlay will be slightly higher – but not always as much as you think. To be able to register any piglets, you will have to join the appropriate registration body. All BPA-registered pigs are eligible to be shown and it's a great way to meet other piggy people, continue learning and have great weekends away. Seriously, pig people should be a barometer of 'niceness'; we have previously dabbled in showing other livestock where we found a lot of the people cliquey, snippy and even vile on occasion. The other competitors were the single reason we stopped showing them. They were long miserable days that cost a lot of money. Our experience on the pig-showing circuit has been completely different. Yes, you get the odd one or two you wouldn't elbow your way through a crowd to talk to, but in the main they are extremely friendly, humorous and helpful.

When buying registered pigs it is a good idea to have a look at the annual breed surveys for your chosen breed to see which bloodlines are the rarest. We decided to go for the rarer British Saddleback bloodlines to give us an added interest, but it does depend upon what you want to do with the progeny when born. If you need to supply a thriving hog roast business, then perhaps the more prolific lines are the way to go, as rare lines might be rare for a reason. They could take longer to conceive, have smaller litters, have weaker litters with lower survival rates, or even have a genetic trait of squashing their piglets. Invariably, the bloodlines you want are miles away, especially in the USA, but you might be lucky. We have been as far as Scotland and Wales to source our rare British Saddleback lines but we have a good feeling within us that we are helping to preserve the rarer lines. That said, all traditional and heritage breeds are still in the rare category with the Rare Breed Survival Trust/the Livestock Conservancy and care must be taken not to overlook the more prolific lines within breeds so that they then become the rare lines.

BRINGING YOUR PIGS HOME

Whether you are buying pure bred, cross-breed, fattening weaners or registered pigs, you need to get them home. It seems obvious, but arrange to collect your pigs when you have time to keep an eye on them. If you haven't taken a week's annual leave, then ideally pick them up on a Friday evening so they can be unloaded straight into the ark for the night, or early Saturday morning so you get two full days of watching them.

If you are transporting a pig older than 10 to 12 weeks, you will need to take a suitable livestock trailer or lorry, which must be disinfected before use and within 24 hours after use or before next use – whichever is the sooner. Always cover the floor with straw and, when loading, place some straw on the ramp to minimise any reluctance to load. Piglets of 8 to 12 weeks (depends upon breed size) can be transported within a secure container/dog crate in your car or van, but careful with vans if you cannot open a window or have an extractor fan on, especially in the summer, or you could have a dead pig at the other end.

A large, robust dog crate with a solid floor (on top of plastic sheeting) full of straw would be considered suitable. The welfare rules on ventilation, ease of cleaning, space to stand and lie down in their own space still apply, but the piglets may be carried in. Don't be alarmed if the breeder picks up your piglets by the back legs and carries them upside down; this actually stops the piglets screaming as much and seems to make them docile (Figure 4.28). With the British Saddleback breed, we only carry them like this below eight weeks of age and always support the piglets' hock joints and upper legs. There is a danger of giving your pigs a hernia if they are too heavy.

In the UK, the breeder should apply a temporary mark to piglets less than 12 months old if they are unregistered. If they are registered, they should show you the ear tag plus either ear notch or tattoo (see Chapter Seven), which confirms you have purchased the correct pig(s). In the USA, you should also expect to be given a litter certificate or birth notification for each piglet that matches the ID of the pigs. Breeders will often deliver pigs and piglets free of charge locally, or charge a

FIGURE 4.28 Neil correctly holding a young pig upside down.

nominal mileage rate for slightly further afield. However, you will need to sort out a legal trailer before your pigs are ready for slaughter.

Unloading and Settling Your Pigs

So you have just taken possession of your first pigs and, as you had everything ready before you left (!), you just need to unload your piglets into their new home. If they are under 8 to 12 weeks (depending upon breed), you can carry them in or place the opened crate inside their pen. A few food pellets just outside the cage may entice them to come out – but let them take their time. This goes for trailers and older pigs as well. Place in the pen entrance, blocking any escape routes if needed, place some straw on the ramp and gently encourage them to come out. Do not leave the trailer in the pen once they have come out or they will eat the tyres and electrics. For the first 24 hours, apart from feeding and topping up water, it is best to let them settle and find their trotters. Older pigs that have been well handled or have been shown can often be moved to their new pen with a stick and board, which makes life easier if the pen is not accessible by the trailer. Younger pigs can be carried in the crate or in your arms, but not by the back legs for any length of time. They will probably check the pen perimeter so they know where everything is and generally have a good nose and scratch. If they are not used to electric fencing, then expect some loud squeals, but no escapees, as you used stock fencing as well. Make sure at bedtime that all pigs are happy in their ark and no one has been left outside, especially if the weather is not brilliant.

We now always unload young piglets directly into the ark, and even if they run out at Mach 4 immediately, they seem to know where it is and go to bed at night time. This can save a fair bit of time the first night and the pigs will not associate bedtime with being scared from being chased. Older pigs are a bit savvier and just go to bed. Until the weaners are used to their new home, put a container of water just outside the ark. That way, if they are thirsty, they are more likely to sneak a drink. You may find that pigs seem to go off their feed for a few days after arrival and you will wonder if you are feeding too much? They are ill? They have the Ebola virus? It is unlikely that they are unwell. It's more likely they have eaten too much of the nice grass in their new pen and are just full, but keep an eye on them.

The time weaners move home is one of the most stressful times in a piglet's life – mentally and immunologically. They have just been weaned and are adjusting to a no-milk diet, they have been put in a crate or trailer for what must seem like ages, and then they are put in a strange environment with people they don't know grinning at them and talking in weird voices. It is no surprise then to expect a bit of scour (loose faeces). It is also another reason piglets should not be plucked off the teat at sale time with no prior adjustment to their non-milk diet in unfamiliar surroundings. It still amazes me that anyone would do that. A decent breeder wouldn't. Ideally, keep them in the group they are used to and don't mix with pigs they don't know until they have settled. Also, try and keep visitors to a minimum for the first few days, especially the ones with precocious, loud or undisciplined children.

NOTE: In the UK, your farm is now on a 20-day movement standstill for pigs.

GET YOUR PIG HEALTH RIGHT

It is said by a fair few commercial pig breeders that middle-class hobby farmers are the biggest risk to the farming industry, and to a certain extent they are correct – although I must point out that the UK Foot and Mouth outbreak of 2001, which cost a total loss of £9 billion and caused thousands of animals to be destroyed, was caused by a commercial pig farm feeding illegal food waste to their pigs.

I am semi-agreeing with the commercial farmers' claim, but they do cast the blanket wide and judge us as all the same. Just as we know little about how they run their farms, they also know very little about how we run ours. I have been on a couple of well-run commercial pig farms and saw a few things that I would never do, such as leaving food trolleys uncovered within the barns, old feed spilled either side of their feeding troughs, farmers not following their own biosecurity display notices, and some walking to and from the farm shop without going through footbaths. I'm not saying the farms were poorly managed, because the pigs were in fine health – but people in glass houses and all that!

There are smallholders who take the health and welfare of their pigs incredibly seriously and I believe are less risk to the national herds than claimed. There are some practices that need tightening up, one of the most important being that the commercial pigs are often monitored for certain organisms and diseases as a part of organised health schemes and treated accordingly, whereas smallholder pigs tend to only be treated when they become ill. There are those who are completely irresponsible (but amazingly, legal) and keep their pet pigs indoors or allow them access to indoors, usually in the kitchen where scraps fall on the floor. These pigs are unlikely to be in the best health; however, the risk to other pigs is low, as they are usually contained. I doubt the use of the UK pet pig walking licence is that prevalent once the pig becomes older and stronger.

At the bottom end of the scale you have the completely irresponsible smallholder who is looking for maximum profit, feeding their pigs very little proprietary feed, providing them mainly with waste vegetables, cake, dairy and bread from various sources. I also have concerns that if profit is the driving force, then what are the chances of them following the unseen rules such as not feeding anything that has passed through a kitchen? I doubt they deworm very regularly, if at all, and they are highly unlikely to vaccinate. Biosecurity is non-existent and calling a vet for advice would be a last resort – by which time they have probably allowed anything infectious to spread. Their pigs lead very sad lives, reproducing continuously with no regard for body condition, and the offspring reared enter the food chain, contaminating the slaughterhouse and butchery *en route*. It is these smallholders that are the primary risk to the national herds – and to decent smallholders' pigs. It is these piglets you may see advertised at cheaper prices or off-loaded at some markets, so think before you buy them and be prepared for extra expense if you do.

An increasing number of small-scale farmers are taking their herd health seriously and are much less of a risk, especially those following strict farm health plans and who make biosecurity a priority. Some smallholders are now addressing the lack of health monitoring and are paying good money to complete health plans. Our herd has a health plan drawn up by our veterinary practice and subsidised considerably by the National Farmers Union (NFU). I would say the majority of smallholdings still have varying degrees of risk, with little or no biosecurity procedures in place, usually through a lack of awareness or sense of embarrassment in asking people to disinfect their boots. These smallholders just let visitors turn up from others farms with muddy boots and buy pigs without quarantining from the rest of their herd or performing health checks. Some hire in boars, which go from smallholding to smallholding, because they don't want to own one, or they take in visiting sows for the boar. Often they don't disinfect their car tyres after visiting the agricultural merchants, they don't disinfect arks regularly between batches of pigs, and they have poor vermin control.

Buying piglets from a reputable source and observing high health and biosecurity procedures means you will have to take less time off work unexpectedly to nurse sick pigs, or to be there when the vet turns up. It really is as simple as that. Often people do nothing about biosecurity until they have had a problem, and then they start thinking about it. That is one expensive lesson to learn, to both the pigs and your pocket.

Keeping your herd healthy and employing the very best standards of stockmanship is essential. You will be rewarded with lower veterinary bills, healthier litters of piglets and better-tasting meat, which you will be able to eat guilt free. Some people have a natural 'stockman's eye' – the ability to spot when something is wrong with a pig, preferably before the pig even shows you that it is ill – while others have to learn how to develop one. People know when there is something not quite right with their children/dog/cat, and that is all a stockman's eye is. The easiest way to develop your 'eye' is to get to know your animals intimately; know their foibles and quirks, as what is normal for one pig may be very abnormal for another. There are no short cuts, so spend time interacting with them, take a coffee up the field and a garden chair and just sit and observe. It is a fantastic way to spend a few hours.

By law you must adhere to the Five Freedoms mentioned in Chapter Three on husbandry. Remember these are minimum guidelines, and exceeding them where possible is an excellent start to keeping your pigs healthy.

BIOSECURITY

To me, biosecurity means the procedures followed to help reduce the risk of transmission of infectious diseases, parasites and pests to my livestock. Time spent on biosecurity minimises the incidence of disease and illness you will come across in your herd. All biosecurity is centered on good hygiene, prevention and common sense. The closer you adhere to your biosecurity plan and observe your herd health controls, then the more likely you are to succeed.

I am repeating myself here, but if you get the husbandry and biosecurity right, then any bug, no matter how pathogenic, has less of a chance of taking hold. If it does manage to get into your herd, then your newly acquired 'stockman's eye' comes into play and prompt treatment or action, e.g. isolating the sick animal, will help prevent infectious spread through your herd.

The most basic biosecurity consists of:

> **TIP**
> It takes no extra time when feeding your pigs to look at each one individually and check that all looks well.

- Keeping all equipment clean and disinfecting if required.
- Keeping equipment used on isolated animals separate.
- Isolating all incoming stock (new animals or show returns) for at least 20 days and keeping them under strict observation for any signs of disease.
- Having separate boots for visiting isolated animals or scrubbing with a Defra- or federally approved disinfectant, e.g. Virkon S™.
- Keeping good stock and husbandry records. You will not remember everything you have done and when.
- Maintaining vermin and pest control, especially in food storage areas.
- Never feeding mouldy or out-of-date feed. I will even return a sack of feed when delivered if it has a hole in it.
- Never feeding kitchen waste to the pigs.
- Always having a stock of disinfectant to hand. If a major outbreak of a disease occurs, it sells out very quickly.

If a disease outbreak occurs, keep visitors to your animals to a minimum, especially if they have their own farmed animals. Make sure any visitors scrub their boots with a Defra-/federally approved disinfectant and change their clothes. Also, have all feed/straw/other deliveries at the entrance to your farm, including your own vehicle, and nowhere near the animals. It means more work for you in the short term, but less work in the long run and it may even save your herd from slaughter!

How Long Should You Rest Paddocks Between Batches of Pigs?

I just knew you would ask me that. I have spent a lot of time trying to research this as we have paddocks that are in all but constant use and I have only come up with a partial answer. The primary paddocks of concern to me are our isolation and our farrowing paddocks. The isolation paddocks house new pigs and those returning from shows, and are more likely to harbour pathogenic bugs. Our farrowing paddocks are less likely to contain bugs, but will be housing the most susceptible to infection . . . the piglets. Viruses are the least likely to survive in paddocks and will be highly susceptible to the elements, so a couple of weeks should see your paddock free from viruses.

Bacteria, particularly spore-forming types, and parasites can live for years in the soil. I did find documented evidence that the eggs of the *Ascarid* worm, a common outdoor pig parasite, can live for up to seven years in soil. Resting paddocks for seven years would be impossible for most of us, and so I looked into disinfecting mud. You can't just pour on the usual disinfectants, however, and most are rapidly inactivated by organic material such as soil and faeces.

I did find the manufacturer of a lime-based outdoor/soil disinfectant and approached them to perform a case study on the farm, with before and after micro-biological analysis (Figure 5.1). It only works when there is no vegetative material, so if you do it promptly after the pigs have vacated, then I doubt there is much about. You dampen the soil if it is dry and then sprinkle the powder at a rate of 500 g per m². The product then heats the soil to approximately 70°C (158°F) over a 24-hour period, and leaves it highly alkaline for the next couple of months. This combination kills most viruses, bacteria and parasite eggs, effectively sterilising the paddock. The beauty of this product is that it is safe to return the animals after 24 hours and so hardly disrupts your husbandry. During our case study, we returned a batch of meat weaners after 24 hours to the treated isolation paddock with no ill-effect to them, and so, feeling braver, opened up a farrowing paddock onto treated ground with very young piglets, and again there was no ill-effect. The microbiological analysis performed was restricted to bacteria and did show a statistically significant reduction in bacteria in both of the pens, but not a complete elimination. Anecdotally, the isolation paddock had previously housed a couple of unwell pigs over two batches, and the meat weaners did not succumb to anything!

FIGURE 5.1 Neil applying a lime-based disinfectant to an outdoor paddock.

Farms with a High Herd-Health Status

You may have heard about pig farms with a 'high herd-health' status. This is where, in addition to efficient basic husbandry, further biosecurity measures are implemented and strictly adhered to. These can range from a few additional measures to absolute paranoia, but I can completely understand why. If you have hundreds or even thousands of pigs all kept in close proximity to each other and disease gets in, it will spread very quickly and cost you a lot of money, and, if it's a notifiable disease, the taxpayer as well.

We probably fall somewhere in the middle, and now we are a partially closed herd, we can start to tighten up certain areas. A closed herd is when you no longer buy animals in, but breed your own future breeding stock, or you buy from another approved farm with a high herd-health status. We have certain areas where we know we need to improve and some areas where improvement would be nice, but almost impossible.

Having said that, we don't need to become a prison, as our pigs are not kept in small areas on top of each other – and pigs do have immune systems, after all.

Our pigs are reared outdoors from birth and so bird control is quite a problem. Birds spread disease with their faeces and mechanically from farm to farm and it is impossible to stop them flying over the pens or landing in them. To help minimise the risks, you can site water troughs away from fences, so that they don't defecate into them when perching, hammer a nail on to the top of posts to prevent perching, feed in the dark to stop them stealing food, and/or shoot the birds.

We sometimes have problems with other vermin being attracted to the animal feed. We keep all our feed stocks and open sacks in vermin-proof containers but it is probably our chicken feed that attracts the rats most, as they are so messy when they eat. We have feral stable cats that have really helped keep the buildings vermin free and we have a poisoning programme that keeps the numbers down. Sadly, some of our neighbours do not, and so we have to keep this up continuously, which is expensive.

We also have a public footpath across our land and some of the public seem to think it is their right to go where they like, which is a problem when they have been traipsing across other farmland. Signs make little difference, except the one which says 'Beware of the snakes' (we don't have snakes, but I was thinking outside the box), so our pigs are sited and fenced off from the footpath and we make sure we shut the gate to the pigs' area after we leave, to help deter them from getting too close. Despite this, we have had people climbing gates with their loose dogs to get a better look and they can be quite defensive and even rude when challenged. When I get cross about it now I intermittently hang a sign on the gate that says, 'Neil, don't forget the pigs in pens 3 and 6 have been treated, so don't touch them or let the dogs near them.' That makes people too wary to let themselves in!

We attend national shows with our pigs and so from that point of view our herd is not closed and we have to treat the pigs as new arrivals when they come back to the farm and isolate them accordingly. Our show and breeding stock pigs are all vaccinated against erysipelas and porcine parvo virus. During the isolation period on our return we also bathe them with Hibiscrub™ and worm them before mixing back in with our main herd. Showing is probably our biggest biosecurity risk. When back on the farm, we attend to all the other pigs first and do the isolated stock last. We also disinfect our boots or use dedicated footwear to do these pigs for the full isolation period. The other risky activity we perform is buying in new boar lines and, again, these pigs are scrubbed, dewormed and isolated before mixing with the main herd.

We also ask all visitors who keep livestock or regularly walk through farmland to wear clean boots and then disinfect their boots with an approved disinfectant before they go up to the field.

In the commercial pig world, the following biosecurity measures would not be uncommon to see, especially in the high herd health units:

- Visitors to the farm are kept to a minimum, and if the visitor has been near pigs, at least 48 hours pig-free must pass. With absolute paranoia, no visitors who have been near pigs would be allowed admittance.
- Spare boots and overalls are kept on-farm for visitors to use and they are often required to shower on the farm and wear protection over their hair.
- Disinfection footbaths must be used between batches of pigs and between pig houses and separate feed trolleys, mucking out equipment, wheelbarrows, etc. for each pig house.
- Pigs are attended to in a specific order starting with the most susceptible pigs through to the least susceptible to disease or dedicated staff for each pig group.
- A strict vaccination programme in place, based upon microbiological testing.
- There is regular monitoring of the herd's disease status by laboratory testing.
- Vehicles are never allowed onto the farm and separate access for feed deliveries and staff vehicles is essential. Disinfection of wheels and bodies of the vehicles is carried out before entry.
- Most farms operate a closed herd system where all pigs are bred from one source or on the farm itself and any new pigs purchased are isolated for a minimum of six weeks in strict quarantine away from all the other pigs, with separate dedicated staff.
- Artificial insemination is used rather than buying in new boars, or purchased boars are isolated for a minimum of six weeks and come from another high-health-status herd.
- Housing is sealed and so can remain vermin free, including birds, and is often ventilated, with the recycled air passing through filters.
- Muck heap or slurry management is operated and all waste is carried away to tanks far from the pigs.
- An 'all in, all out' system is operated, where pigs are kept in one pen from weaning through to a certain weight or size and then this pen is thoroughly cleaned, disinfected and even steam-sterilised before the next batch of pigs is put in.
- Pressure is put on anyone keeping pigs within a two-mile radius of the commercial unit to cease the activity – which can be a particular problem if you are renting the land.

A PIG'S ANATOMY

To assist in understanding how the health of the pig is linked to many different bodily functions and how some of our veterinary and technical interventions work, I have provided a brief overview of a pig's anatomy. The anatomy and physiology of the pig can be broadly grouped into 11 highly inter-related systems: circulatory, digestive, endocrine (hormone), immune, muscular, nervous, reproductive, respiratory, sensory, skeletal and urinary. A meaningful section, which didn't induce immediate restful slumber, on all 11 would be impossible; indeed, whole books have been written on each individual system. So I have taken a holistic view of a pig's anatomy and tried to relate it to familiar parts of the pig that we already know about.

Let's start with the skeleton of the pig, the function of which is multi-purpose: it provides a frame against which the muscles work to provide movement; it offers support and protection for the internal organs; it enables blood cell formation from the bone marrow; and it provides an effective calcium storage system. A skeleton is, as we know, made up from the sequential joining of bones, some fixed (e.g. the skull), some semi-movable (vertebrate) and some freely movable (leg). Bones are incredibly strong and filled with a spongy tissue mass known as bone marrow and are protectively encased in a membrane made up of connective tissue. Moving limbs consist of bones that are linked together through joints, which consist of two ends of bone held together by ligaments and muscles with smooth, shock-absorbing cartilage positioned at points of friction, which prevents the bones from touching and damaging each other. To assist in friction-less movement, each joint is bathed internally with a lubricating fluid. The muscles and ligaments attach to the bones via the connective tissue membrane, effectively holding the bones and joints together. Near the ends of the bones before each joint are flat-tened areas of cartilage running at right angles to the bone – the epiphyseal or growth plates. By increasing their thickness, they cause bones to grow in length and width from birth until maturity. The separation of bones at these plates is a common occurrence, especially in young fast-growing breeds or overweight growing pigs, causing leg weak-ness and lameness. This is one potential cause of the more serious necrotic condition osteochondrosis, and can be a serious problem in working boars. Bone is continually being renewed, even in adult pigs that have stopped increasing in size, meaning they are able to repair fractures and respond to pressures. The main pressure on bones is from fitness/muscle tone, which can only be enhanced by exercise. Pigs that are able to exercise are likely to have stronger bones and joints than those that can't, and pigs kept in stalls or crates have softer more brittle bones than sows kept in larger pens or outdoors, especially during late pregnancy and lactation.

Surrounding the skeleton is the muscle – also known as the meat or flesh. This is more familiar territory to the meat-producing pig keeper and we have all studied dia-grams of the different meat joints and from where they originate, even if it's only so we don't look too stupid at the butcher's.

There are two types of muscle – involuntary (smooth and cardiac) and voluntary (skeletal) – and although both may be eaten, it is the larger, skeletal muscles that pro-vide our picnic hams and chops. The smooth muscle is found in the walls of hollow organs, e.g. intestines and stomach. They work automatically and are involved in many 'housekeeping' functions of the body. The muscular walls of the intestines contract in a wave to push food through the body. In the bladder, they contract to expel urine from the body, and there is even one in the eye – the pupillary sphincter muscle – which shrinks the size of the pupil in response to bright light.

The cardiac muscle is, of course, only found in the heart and uniquely never gets tired. The skeletal muscles are there to maintain posture, allow voluntary movement, stabilise joints and generate heat. It is also these muscles that can get damaged through external trauma or pressure to cause lameness, bruising, swelling through blunt or sharp tissue damage, and abscesses through infected puncture marks and wounds.

In terms of eating, not all muscle seems to be equal, despite numerous books promoting nose-to-tail eating or eating everything but the squeak. In the UK, it is said that if you draw a straight line from the top of the pig's head to just above the rear hock joint, then all the meat above the line can be considered premium cuts and all the meat below the line inexpensive cuts (Figure 5.2). So the spare rib roast (my favourite), top of the shoulder, tenderloin, loin chops, chump chops and hams are all considered premium, leaving the belly – including the thick end and rack of spare ribs, trotters, hand (top of foreleg/lower shoulder), chaps and head. I have used the meat names for the muscle rather than their 'proper' names to make for easier reading.

Interestingly, the Americans draw two horizontal lines, one along the top of the legs and another from the bottom of the ear straight through the middle of the body and then veering up to the top of the tail, dividing the pig in three – high on the hog, middle of the hog and low on the hog (Figure 5.3). Here the middle of the hog is the primary meat for sausages.

Protectively covering all of the skeleton and muscle is the pig's skin. Apart from turning into delicious crackling when cooked (seriously, Americans, you are missing out on a treat when you de-skin), the skin is a major organ and it has many functions that keep the body in homeostasis (balance). It provides protection from the weather as skin is waterproof, is a barrier to pathogenic organisms in the environment, enables the regulation of body temperature from the covering of hair to the evaporation of excess heat from wallowing, supports the balance of fluid requirements and houses the pig's sensory endings such as touch, temperature, pressure and pain. With the addition of the pigment melanin in varying amounts, the skin can take on different colours from pale yellow to black, assisting in camouflage (think striped piglets) and even in fertility rates. With additional keratin, the skin hardens to make hooves, which protect the whole structure of the pig from the ground. The skin also has the ability to convert ultraviolet light into vitamin D, an essential component of strong healthy bones, and a prime example of how the anatomical systems are linked.

Powering all of the pig's functions is the digestive system. The contents of the entire tract are independent to the pig and are pillaged for useful nutrients (vitamins, amino-acids, fats) and water as they pass through before the indigestible waste is expelled. As food is collected in the mouth, the teeth crush the food to maximise the accessible surface area for the enzymes in the saliva to begin breaking down the components. Once

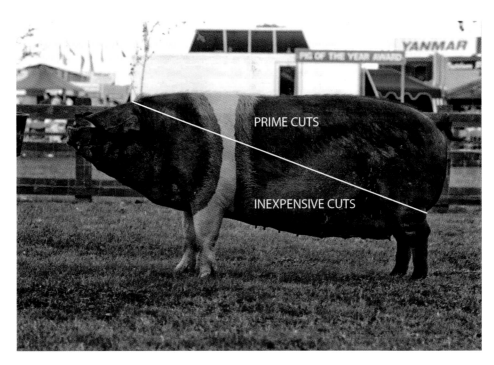

FIGURE 5.2 The division between premium cuts and inexpensive cuts on the pig (UK).

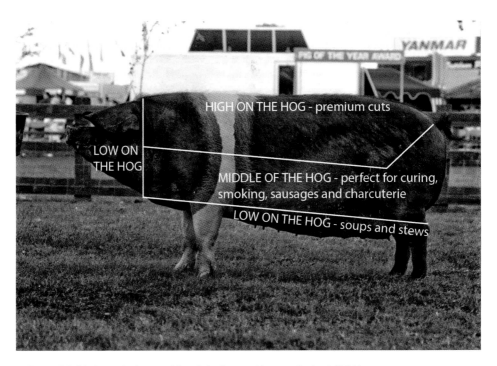

FIGURE 5.3 'High on the hog, middle of the hog and low on the hog' (USA).

the food is crushed, the back of the mouth opens into the pharynx, which is the communal area for the passage of both food and air. A flap of tissue called the soft palate automatically moves to protect the opening into the windpipe when swallowing, allowing the food to proceed through the oesophagus and down to a single (monogastric) stomach.

The food continues to break down into components, releasing nutrients in an accessible format for digestion as it passes through the stomach and into the small intestine. In the pig, this includes both vegetable and animal material, as the pig is most definitely an omnivore, despite the current UK/USA ban on feeding meat products to pigs.

The intestine has two distinctive parts, the small and the large intestine. The small intestine in cross-section contains millions of finger-like projections called villi. These villi increase the surface area to maximise the amount of contact between the desired nutrients and the digested food passing through. It is damage to these villi by bacteria, viruses or parasites that contributes to weight loss in unwell pigs by disrupting the efficiency of the digestive process. The large intestine or colon commences with the caecum, a pocket within the intestinal tract responsible for the friendly microbial digestion of cellulose, which is found in plant material, including grass. This organ in the human, present in a similar although vestigial form (and called the appendix) is now redundant, and so we do differ slightly.

The colon is primarily responsible for the reabsorption of water, vitamins and electrolytes from the mixture of food, saliva, and gastric juices passing through. This prevents excessive water loss and therefore dehydration and is the reason that pigs can be effectively rehydrated via the rectum through the flutter valve technique described later in this chapter. The dried waste is then excreted via the rectum and anal sphincters.

While this information isn't remotely likely to help you complete and pass an anatomy exam, I hope it has given you a modest insight into some of the functions within the pig and how keeping a pig healthy can potentially affect every part of the body.

ROUTINE HEALTH PROCEDURES

There are some routine jobs, which contribute to the well-being of your pigs, that you must know how to tackle.

Taking a Pig's Temperature

In truth it is rarely necessary to check temperature, but if an animal is off-colour and you wish to see if she is running a temperature, then any veterinary rectal thermometer will suffice, although a digital one is easier (Figure 5.4). The thermometer is gently inserted into the rectum, taking care not to hurt the pig – ideally along the inside of the rectum wall. In digital thermometers, you will often hear a beep and then you can read

FIGURE 5.4 How to take a pig's temperature.

Photo courtesy of Practical Pigs magazine.

off the temperature. The normal body temp is from 38.5 to 38.8°C (101.5 to 101.8°F), although young piglets run about 0.5°C (0.9°F) above this and sows in season run a bit higher as well. Some form of restraint may be required. Piglets can be held or older pigs might stand for food if well enough to want to eat, or try containing them inside a triangle of hurdles with people standing on each one so they can't be lifted.

Injecting

The purpose of medication is to benefit the animal, and incorrect administration can potentially do serious harm or just not work and so be of no benefit. The route of administration is dependent upon the medicine. Medication can

FIGURE 5.5 An injection being given behind the ear.

be given orally in feed or water, or via the mucosal membranes of the snout or rectum, as a topical application – straight onto the skin, or directly into the body as an injection. If you progress further than buying in fattening weaners, then injecting is a technique that you will have to learn at some point. In the pig, injections are most often linked to deworming, vaccination or medication that will have been prescribed by a veterinary surgeon (or an SQP for routine products such as dewormers and vaccines) for use on your pigs only. Be aware that borrowing from another pig keeper is illegal.

There are approximately 14 different routes of injecting, although most require Member of the Royal College of Veterinary Surgeons (MRCVS) or US State Veterinary registration to perform them. The most likely ones you would use as a part of your routine husbandry are subcutaneous (SC or SQ), so just under the skin, or intra-muscular (IM), into the muscle or meat of the animal. Injecting is not a job for the timid, and even unwell pigs can put on a spritely run when you try to inject them, leaving the needle and syringe bouncing off into the distance.

Before using any product, read the label. By law, all medicines have product information sheets that list everything you need to know, including the storage conditions, expiry dates after opening, target species, routes of administration, dose rate for different ages/weights, and withdrawal times – the time that must be left between giving a medicine and allowing the animal to enter the food chain.

Vaccines are usually a standard dose size for all ages but most medicines vary in the quantity you give according to the weight of the pig, so it's imperative that you know the live weight of your pig so you don't under- or overdose. If a piglet is small enough to be held, you could hold it whilst standing on a bathroom scales. You could use a purpose-made weigh tape to give a reasonable ballpark figure, but just be careful that it is not a deadweight tape or you will be underdosing. Sheep- or cattle-weighing scales can also be used.

As mentioned earlier, the pig will need restraining in some way. Food can be excellent at keeping the pig still long enough to perform a quick, low-volume single injection and, if you are proficient, it can be all the distraction you need. If you have a sheep-weighing crate or a robust and secure small area, then perfect. If you need to perform an injection where you need to give multiple shots due to the size of the dose being too great for one injection site, a pig snare is useful. These fit around the top jaw of the snout and effectively secure the pig, but you will require a second person to do the injecting.

If you have numerous pigs, it may be worth investing in a multi-injection or air pressure injection gun, which will do exactly what the name suggests, so even if they do run off, you are left holding the gun. Once jabbed, twice shy and it makes sense to cause as minimal pain and stress as possible, so consider investing in equipment such as the Slapshot™, Masterject™ or similar, to make life easier for all. The Slapshot™ is effectively a long piece of tubing with your syringe (or gun) at one end and your needle at the other, and it is the best money I have ever spent. As you 'slap' the needle into the muscle, you can follow if the pig moves away, whilst continuing to press the syringe and then pulling on the tubing to retrieve the needle. The advantage of this system is that continued pain doesn't occur, as the needle isn't moving about, tearing muscle tissue inside the pig. The Masterject™ is a semi-automated syringe on the end of a long pole which allows you to inject from a safe distance or even from horseback, as it says on their website. Each to their own – that bit isn't compulsory.

Choose a clean area of dry skin and follow the instructions on the product leaflet. The best site for a SC injection is approximately 3 to 7 cm behind and slightly above the base of the ear, using the needle at a 35 to 45° angle in a growing or mature pig – or inside the thigh under the fold of the skin on a tiny piglet (Figure 5.5). It can help to pull out the softer, loose skin, forming a tent before injecting. For IM, behind the ear is also the preferred site, although you must choose a needle length that will pass through the fat layer and into the muscle. This time, the needle enters at a 90° angle, i.e. perpendicular to the skin. The only exception is when IM injecting very young piglets, e.g. for iron injections. The preferred site is then the larger thigh muscle. Injecting behind the ear has an added advantage, as if you do cause an abscess or scarring to form, only a small, less valuable part of the carcass will be condemned at the abattoir.

Keep everything aseptic (germ free) by washing your hands, using alcohol wipes on the bottle top, and only ever put a sterile needle into the bottle to draw the medicine out and remove when finished. Try and use sterile, disposable needles and syringes that you can discard after a single use. If you do need to use reusable extra-strong needles, e.g. on a senior boar or when using a Masterject™, flush with boiled water and sterilise before next use. Don't use detergents, as any residue may render a live vaccine useless. This is also applicable for multiple dosing guns, which are designed to withstand the cleaning process.

When choosing what needle size to use, pick the smallest diameter (highest gauge (G) number), to allow the viscosity of the product to easily pass through. This minimises any tissue damage or leakage of product from the injection site. If the drug has a thick consistency and/or you need to get a large volume into the pig, then a tiny, fine needle would be useless. Select a needle length for the type of injection, e.g. for SC injections it only has to be long enough to get through the skin layer. For IM injections, it must be long enough to get through the layer of fat, which will be highly dependent upon breed of pig and how much fat it has been allowed to lay down. Be warned, though, that too long a needle could cause unnecessary tissue or nerve damage.

Some people use the same needle within batches of pigs when doing routine tasks, such as deworming or vaccinating, but I would still recommend changing the needle every five or so pigs and definitely don't practise this if you are treating the pigs for an illness.

The volume of product required is solely dependent upon the weight of an animal and the concentration of the medicine and there will be a maximum volume that can be given at one site or it may need performing daily for a few days. Try and choose subsequent sites approximately 4 cm from the original site or use the opposite side of the pig.

Once you have your pig restrained, take the carefully mixed bottle of medicine or product to a clean area nearby and then open your sterile syringe and needle. Draw the same volume of air into the syringe as you need of the product and inject the air into the bottle. Holding the bottle upside down, immediately pull on the syringe and fill to the required volume. Withdraw the needle and hold the syringe needle up and tap on the barrel to get any air bubbles to the top and carefully squeeze the air out by pushing slowly on the plunger. Now you can either insert the needle into the pig first and then attach the syringe, or insert the needle already attached to the syringe as one unit. To prevent leakage after IM injections, you can pull the skin to one side before inserting the needle, so when the needle is withdrawn, the holes don't match up. Either way, for IM injections, pull back on the syringe and make sure you haven't hit a vein, as injecting a product not designed to be intravenous (into a vein) can have disastrous consequences. For SC injections, this is highly unlikely, so just form your 'tent' of loose skin or inject in at an angle. If you are using a Slapshot™ or Masterject™-type system, follow the manufacturer's instructions.

Always know what the potential side effects might be; keep a close eye out for them and know what to do if you see them. You have created a puncture wound that could potentially form an abscess and if this occurs you will need to deal with it, but if you follow the instructions given here, then all should be well. As always, if you are not comfortable or competent in injecting your animal, ask someone more experienced or your vet for some advice and training.

TABLE 5.1 A guide to needle lengths used at the different live weights. If possible always choose the smallest gauge.

LIVE WEIGHT OF PIG	METHOD	NEEDLE LENGTH AND GAUGE		METHOD	NEEDLE LENGTH AND GAUGE	
Less than 7 kg		5/8″	21G		5/8″	21G
7 to 25 kg		1″	19G		5/8″	21G
25 kg to 60 kg	Intramuscular (IM)	1″	19G	Subcutaneous (SC or SQ)	1/2″	19G
60 kg to 100 kg		1.5″	16G		1/2″	19G
100 kg+		1.5+″	16G		1″	18G

NOTE: If you have lost the product information sheet, then look up the product name on the NOAH Compendium website (www.noahcompendium.co.uk) or the Compendium of Veterinary Products (https://valleyvet.naccvp.com) and print off another copy.

Vaccination

Vaccination in the outdoor-reared smallholder pig is less commonly practised except for a few key vaccines. There are commercial vaccines available for a number of diseases but most are applicable to intensively reared pigs all living in very close proximity to each other, or if you have had a problem diagnosed on the farm. The vaccines that are the most useful to the outdoor breeding pig are against erysipelas and porcine parvo virus (PPV). Erysipelas can cause abortion and foetal death, so vaccination is important to future breeding stock. PPV is the most common and important cause of infectious infertility. Most smallholders only vaccinate their breeding stock and not their fattening weaners. Some breeders now vaccinate all their piglets against certain bugs that affect the young. A common one is against a mycoplasma that causes pneumonia, and this may be a good call when you have lots of piglets being born or you are exhibiting sows and litters. There are many vaccines for many diseases and we would consider different ones if our laboratory monitoring identified a potential problem but, at the moment, only our breeding herd and future breeding stock (for sale or keep) are vaccinated. In the USA, there may be vaccines that are required by the state or even federally, or which require permission to administer, e.g. Aujeszky's Disease (pseudorabies), so ask your state veterinary surgeon for further advice.

TIP

Most vaccines are injectable and have to be used within ten hours of opening, so time your vaccinations so that you can use most of/the entire vial the same day.

Deworming Your Pigs

Outdoor-reared pigs have a worm burden even in the best-kept herds; the idea is to keep the burden to an acceptable level to the pigs' health. Pigs are usually dewormed every six months, before farrowing, and aged between four and eight weeks prior to departure for their new home when sold as fattening weaners. Just because your pig looks well doesn't necessarily mean that parasites are not present, multiplying and causing unobservable damage. In meat weaners, this could result in using more food to get them to pork or bacon weight, meaning it would have been cheaper to deworm them anyway.

In breeding stock that hasn't been dewormed it may take longer to reach a weight suitable for mating; gilts may not be able to hold a pregnancy, or they may give birth to weak, or underdeveloped piglets that are at greater risk of dying. It is just not fair on the pigs not to deworm. It is not worth your reputation as a breeder to sell unwormed weaners and you will have weaker stock to sell.

If you have bought your weaner or older stock from a reputable breeder, then your pig should have been dewormed against the most

TIP

Always check when buying fattening weaners that they have been appropriately dewormed. Often, cheaper weaners have not been, as breeders have to keep their costs down somehow.

common parasites. In the fattening weaner, this means that you should not need to worm again before slaughter weight. If you are keeping the weaner beyond slaughter age as breeding stock or pets, or if they are already over six to eight months when you purchased them, then you will need to deworm again. There are numerous products on the market, some injectable and others to be administered orally in the feed or water. Some treat only internal parasites, while some the worm eggs as well; others will also kill some external parasites like mange and lice, so choose appropriately. Annoyingly, dewormers come in pack sizes to treat a lot of pigs, so if you only have a couple, you will not use up the pack before the use-by date. Legally, you cannot give your spare doses to anyone without your vet's permission. Your veterinary surgeon will be able to dispense a smaller amount.

Wormers come in variety of different trade names but may be the same chemical or group of chemicals. Your veterinary surgeon or the Suitably Qualified Person (SQP) at your local agricultural store will be able to help you select the right product.

To worm organically, you may need to double or even triple the withdrawal time of the wormer, so check with the Soil Association/USDA websites for guidance. Do not

TABLE 5.2 Some of the common deworming preparations and their activity against parasites.

Active medicine	Doramectin	Ivermectin 1% and 0.6%	Amitraz 2%	Febantel	Flubendazole	Fenbendazole
Trade name example	Dectomax	Ivomec Virbamec	Topline	Bayverm	Flubenol	Panacur
Administration route	Injection	Injection or in feed	Pour-on	In feed		
Endoparasites (Internal)						
Activity against worm type: Large White / Red stomach / Lung / Nodular / Whip / Thread		YES	NO		YES	
Stomach				NO	YES	
Kidney				YES	NO	YES
Larvae					YES	NO
Eggs	NO				YES	
Ectoparasites						
Lice / Mange mites / Ticks		YES	YES		NO	

TIP

Don't forget to record in your medicine book any medication given, on the day of administration, if it was performed on your farm.

be fooled into thinking natural herb-based 'wormers' are better for the pig and as effective as chemical wormers; they have not yet passed any legitimate clinical trials and have not been proved to work. In my opinion, you might just as well feed your pigs a £20 note or $20 bill a week, as the net result would be the same.

You will note from the table that the dewormers with 'mectin' in their names do not kill the worm eggs but do kill external parasite infestations. This may help you choose your dewormer, as you may wish to kill worm eggs as well.

Killing the worm eggs will effectively reduce the worm burden in your paddocks and may increase the time period before the paddock gets 'pig sick', i.e. has a high parasite burden, or has large amounts of faecal matter in the soil. Interestingly, if you ever have a liver condemned at the abattoir due to 'milk spot' and the pigs are on your paddock for the preceding eight weeks or more, then you have the large white worm *Ascarid* eggs on your paddocks. If you worm your pigs at least eight weeks before slaughter and put them on fresh pasture, then the milk spot evidence in their liver would be gone, but keep an eye on the medicine withdrawal times. As mentioned earlier, the withdrawal period is the time between administering a product or medicine, including dewormers, and when you can legally eat the meat – so, in other words, the slaughter date. You will be asked about any medicines administered on the Food Chain Information (FCI) part of your eAML2. Most dewormers have a withdrawal time, which can range from seven days to a couple of months, so be careful, if the pigs are due to go to the abattoir, that you don't use one with a long withdrawal period if they are booked in a couple of weeks' time.

Natural Remedies vs Licenced Medicines

Licensed medicines and natural/complimentary remedies are big business in the veterinary world but, in reality, what is the difference between them?

I use natural remedies – there I said it – most often as a first line of treatment. If my pigs seem a little bunged up and producing hard stools, I can be seen buying ripe bananas or older, juicy fruits especially for them, to get their bowels going again. If they're a little loose, I administer a few green bananas to help bind them up a bit and if they have what I think might be a viral infection (sniffles, mild cough), then I rush straight to the grocer's for nice fresh pineapple or two, as pineapple contains the enzyme bromelain, which is known to destroy the infectious viral epitopes and is reported to have anti-inflammatory properties.

I have similarly been known to smear a little of our farm's honey on a wound to offer a barrier and natural antiseptic protection. I even once had a case of 'staggers' that I thought was caused by a bacterial toxin blocking the absorption of water within the large intestine. After reading that it was most common after weaning, I assumed that immunoglobulins (antibodies) present in the sow's milk were protective and treated this pig orally with syringes of whole milk; smugly, the pig recovered. Did I effectively treat this pig with a natural remedy? Who knows? And therein lies the issue I have with natural remedies to disease and ailments. With the exception of the use of honey, there is no clinical scientific evidence on the validity that any of my methods work, and the scientific evidence for the use of honey is of 'low quality'. I know that I feel better after I have noticed an ailment and have given my pigs something to 'treat' it, but, in reality, I doubt I'm doing more than doffing my cap to the problem, and the chances are that if the pig recovers, it was probably going to anyway.

The methods I have described so far are 'home made', but what about those that are sold in agricultural stores across the globe? According to the UK Veterinary Medicines Regulations, any product making veterinary medicinal claims needs to be licensed by the Veterinary Medicines Directorate. In the USA the Animal and Plant Health Inspection Service (APHIS) and the Food and Drug Administration (FDA) control the licensing. A licensed product has to *clinically prove* its efficacy and safety for all the animal species it is licensed for use in, using robust clinical trials to pre-defined quality standards. This may also be required per batch produced, to maintain the licence. A similar system is in place for most developed countries.

Products termed 'nutritional supplement' or 'natural remedy' that are not licensed *do not* require the manufacturer to produce any proof or evidence of efficacy and safety. Natural remedies are easy to buy 'off the shelf' because they are not subject to any controls and a veterinary surgeon or Suitably Qualified Person (SQP) is not required in order to purchase them. That is not to say they are dangerous to the individual animal. If adverse reactions did occur and were reported, other laws kick in to prevent the sale. The danger, in my opinion, is when people buy such products, assuming they perform a particular function, such as deworming, and in reality the product does nothing and the parasite levels are building up, often unseen, inside the animal over time.

Livestock keepers can be swayed to think that a product performs a certain function by the use of language such as 'scientifically prepared', 'improves intestinal health' or 'an effective alternative', but rarely does it state what it is an alternative for, how health is improved, or what scientific methods have been used. Other statements such as 'organic system approved' doesn't mean that the product works for what you are being led to believe it does. It just means that it doesn't contain an ingredient that contravenes an organic status – very, very different. Interestingly, for treating pigs within an organic system, the same medicines (perhaps with a few exclusions) are allowed to be used but with longer, adjusted withdrawal times.

There are many 'natural' products available for pigs: diatomaceous earth, herbal intestinal health products, pumpkin seeds, cider vinegar, rosemary, activated charcoal, sugar, garlic flakes, and there is a plethora of anecdotal evidence from livestock keepers, all over the Internet, that state they have worked successfully. What they all lack,

TABLE 5.3 Worm count summaries.

GROUP	MINIMUM	MAXIMUM	MEDIAN
A (Licensed medicine)	0.00	19.00	0.00
B (Herbal treatment)	108.00	1327.00	425.00
C (No treatment)	74.00	958.00	302.00

however, are the results of a clinical trial proving they have worked. I could only find one published clinical trial comparing the efficacy of a licensed medicine (Group A) with an easily purchased herb-based product (Group B) and an untreated group (Group C). The infected animals used were randomly divided into groups and then each group assigned a treatment or no treatment. Each product in Groups A and B was dosed as recommended by the product information and the faeces were examined for the presence of worm eggs from the animals over a two-week period. The results in Table 5.3 clearly show the efficacy of the licensed medicine, with very low worm egg numbers detected. In this trial, giving the animals no treatment at all proved better for the animals than being administered with a herbal preparation.

The use of non-medicinal products is now being scientifically researched, particularly in the pig and poultry industries, and once the results of these research trials are published, they will make for some very interesting reading and hopefully some improved husbandry practices. The use of pro- and pre-biotics and the acidification of water are being explored as a prevention tool, particularly with *Salmonella* in pigs. *Salmonella* is a massive industry problem, and due to the build-up of anti-microbial resistance and the encouragement to use antibiotics less frequently, multi-pronged approaches are now considered essential to reduce on-farm levels to as low as possible.

The single best approach to pig health remains to feed a high-quality, balanced diet, deworm and vaccinate regularly with products that have been proven to work, and keep your levels of husbandry and biosecurity high. If you wish to use fewer chemical-based products for parasite control, then have worm counts on the pigs faeces performed regularly and then use a targeted dewormer as and when you have to. The primary value in natural remedies that I can see is that you increase your observation of your pigs' health and behaviour. The danger comes when you rely upon them without proof or evidence that they are working!

I appreciate this section may be contentious but this is my current view based upon personal experience, research and published claims.

Foot Trimming

The trimming of pigs' hooves is not likely to be required in a couple of fattening weaners you have bought for pork or bacon, but older pigs or those you have kept to run on may require some attention. The most common reasons are that a chunk of the hoof has chipped off or has become overgrown due to being kept indoors on non-abrasive straw. Pigs are not very co-operative about having their feet trimmed, unless you desensitise them to the 'weird' feeling. Luckily, it is not a regular job. We have pigs that are five or six years old and have never required attention to their feet, and if you start to see overgrowth, then avoid the need for trimming by taking them for walks on areas that will wear them down, such as concrete. When it is required, you will be glad that you have got to know your pigs and they are receptive to a good belly rub. During the belly rubs, familiarise the pig with you playing with its legs and hooves and, if you can, get it used to you rubbing a file on the outside edge of the hooves. This will desensitise it to the feel of something on its hooves and may even eliminate the requirement for further trimming. Pig hooves consist of two primary toes/cleys and two dewclaws at the back. They are very hard, and so if attention is required, then spring-loaded side cutters (also called diagonal cut mini pliers) or similar are required to perform the job properly. Clean the pig's hooves of any flaked nail until you reach the smooth layer. Do this on the outside edge only and make sure you leave a blunt not sharp edge. Use a metal file to smooth any sharp edges. Do not cut between the toes on the inside edge. The dewclaws at the back can be shortened if necessary but leave them with a dull cut. Smooth away any rough edges with a metal file. Do one foot at a time and complete the job. This is so that if the pig decides it has had enough and gets up, at least one cley, or one side of one cley, has been done. It may take several attempts to get all four feet trimmed. If the job is urgent and/or faffing about with belly rubs isn't working, you can medically sedate and get the job done quickly. Be careful when sedating boars to follow the maximum dose instructions and not to give too much, as it may cause their penis to extrude, giving rise to the possibility of damage.

Teeth Clipping

Don't be scared into thinking that you will need to attend a pig dentistry course. Young piglets can have their teeth clipped to stop injury to the sow's teats and other piglets' tails and ears. It is a practice more commonly performed on commercial pig farms and it is starting to become more frowned upon even there, as studies have shown that it can continue to cause pain for up to 15 days. It is routinely done 6–24 hours after farrowing in order to prevent damage to the teats while the piglets are fighting for prime position, making the sow reluctant to feed them and to help prevent damage to the piglets when they start play-fighting each other as they get a little older.

We find that in our outdoor-born piglets, which have an enriched environment of mud, tree trunks in the pen, leaves, earthworms, plants, etc. and are not competing for limited space to play, teeth clipping is not necessary. In fact we have never had to perform it yet. If you do find you have this problem, then it could be considered, but make it your last resort and first try looking critically at your husbandry and why it is happening. If you have no choice, then pick up the piglet and hold in one hand. Using the first finger on that same hand push the finger across the back of the jaw to keep the mouth open and the tongue down. You will see four pairs of needle-sharp teeth, and using sharp, clean and preferably sterilised teeth clippers designed for the purpose, place them parallel to the jaw bone and using a quick, firm squeeze cut the top of the teeth to remove the sharp points. Some instructions say to cut to the gum line but most say this is not necessary. Take care not to cut the piglet's gum or tongue. Repeat the procedure for the three remaining pairs. Make sure there are no remaining sharp edges, points or fragments, and tidy up with the clippers if required.

Tusk Removal

If you are going to keep a breeding boar for any length of time, then you may need to consider what to do with his tusks. Boars have four continually growing tusks that can be extremely sharp, although they are usually two to three years old before they become a problem. The upper tusks are often 3–5 inches long, but have sometimes been recorded at 9 inches in length. These upper canines curl up and out along the sides of the mouth.

FIGURE 5.6 An experienced pig keeper detusking a boar.

Photo courtesy of Liz Shankland.

The shorter, lower canines also turn out and curve back toward the eyes. Often one or both tusks are broken or worn from use. The boars use their tusks for defence and to establish a dominance hierarchy during breeding. Due to the injuries inflicted by these tusks, boars naturally develop a thick, tough skin of cartilage and scar tissue around their shoulders for protection. Tusk removal requires the boar to be restrained using a pig snare in most cases. A saw wire (it looks like a thick cheese wire) quickly and painlessly removes the tusks. Ideally get someone experienced to show you the first time so you avoid cutting the boar's mouth or tongue (Figure 5.6). Sedatives, such as Stresnil™, are available to the smallholder on veterinary prescription, but they can be dangerous to use in boars as every muscle relaxes, and if the boar as he gets up treads on his penis, then no non-lethal amount of sedative is going to calm him down, let alone allow you remove his tusks. Clear instructions on the maximum dosage must be adhered to. Other methods include the use of cutters, but supreme care must be taken that the tusk does not shatter, causing more problems than it solves.

NOTE: For safety, boars over 12 months of age must have their tusks removed when exhibiting or showing.

Ear Tagging

Not strictly to do with herd health and more of a legal requirement; however, it does fit in with the Five Freedoms to perform the procedure correctly and appropriately. There are two different types of ear tag – ones that attach to the outside of the ear, which in the UK are usually metal (Figure 5.7), and ones that are applied in the middle of the ear, which are usually a heat-resistant plastic (Figure 5.8). All ear tags are designed for single use only and it is illegal to remove an ear tag in both the UK and the USA once applied.

The most important consideration when tagging is to avoid hitting the two veins that run in a ∧ shape from the base of the ear. They are clearly visible on the inside of the ear, or felt if the ear is a tad hairy. The metal tags are a V-shaped metal strip with a hole at one end and a catch at the other and the applicator pushes and seals them together around the outside ear edge of the pig's ear. They are the easiest to apply as it is impossible for them to get retained in the tagging applicator and so are commonly used to identify slaughter animals. At four months plus of age they do not go anywhere near a vein when applied; however, if they are used to identify younger stock, then a certain amount of space must be left for growth of the ear. This makes them highly attractive to other pigs in the group, which will try and pull at them, often ripping the ear, and the reason why in breeding stock plastic button tags are used. If you use or purchase pigs with metal tags and it becomes a problem, as long as the ears are healed, then rub a bit of mud on the tags to disguise them.

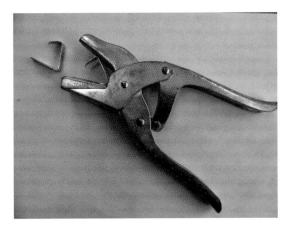

FIGURE 5.7 A metal ear tag and applicator.

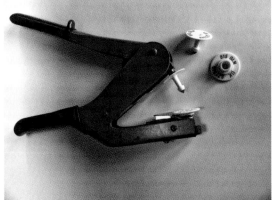

FIGURE 5.8 A plastic ear tag and applicator.

Plastic button tags are two discs of plastic, one with a pin and the other with a hole, which are applied through the middle of the ear between the two veins. Button tags are harder for the other pigs to remove as they lie flat against the ear, although there is a degree of movement between the two discs allowing for growth in the thickness of the ear, which makes them most suitable for breeding or management tags through the pig's life. They do occasionally get retained by the applicator and so you must be ready after squeezing to rectify any retaining issues. We find they most often get caught on the part that holds the pin and a gentle slide off is all that's required. Full instructions are usually supplied with the applicator.

Ear Tattooing

Often used in pedigree identification. The most usual method is via applicator pliers in which you insert the appropriate alpha-numeric tiles (Figure 5.9). The tiles have raised needles with a pointed or chisel end tip in the mirror-image shape of a letter or number. The applicator is squeezed onto the ear with either the tattoo ink applied to the ear or needles or applied dry and the ink is rubbed in afterwards. We now do the latter method as I was fed up trying to scrub ink off the applicator after each use. Ink can be green or black, with the green ink having the most longevity. Tattooing is best performed on young pigs, as with some pliers the ears can get stuck on the needles and need to be gently pried off, with the piglets squealing their heads of and understandably struggling. If you can buy pliers with a special plate to push the ear off, then I strongly advise it. We find this a three-person job: one holding the piglet, a second performing the tattooing and the third to pass the ink/take the pliers. It's a good idea to wear latex or nitrile gloves as it is a very messy procedure.

TIP

Check the tiles have been inserted correctly before use by squeezing down on a piece of paper. It will read how it will look in the pig's ear.

Shoulder Tattooing/Slap-Marking

Most often used for identification at UK abattoirs. This is similar to ear tattooing in that alpha-numeric tiles are used. This time they are larger and placed in a block/holder on the end of a long handle. The tattoo ink is applied to the needles and a firm bang to the pig's shoulder imprints the tattoo into the skin. This is then repeated on the other shoulder. Legally you can only do this once on each shoulder and surprisingly few pigs mind this process, especially if they are eating. This is easily a one-person job.

Ear Notching

Most often used in pedigree identification but occasionally as a management tool. Here chunks are cut out of the pig's ears at specific locations, each representing a number. The location/numbers are explained in the pedigree identification chapter (Chapter Seven). Specialist ear notchers can be purchased from agricultural merchants and online, and come with different-shaped ends – they look like the old bus conductors' ticket punches (Figure 5.10). We use the church-steeple-shaped one as it seems to make a more defined

FIGURE 5.10 A set of ear notchers and a clean set of tattoo pliers.

cut that doesn't make the ear fold in at the notch points. We notch our pigs at around five to six weeks of age just before weaning, with one person holding the piglet and another performing the notching using the whole length of the cutter to make a large notch. If you use small cuts, there is a risk of the ear cartilage moving into the gap, resealing the ear and masking the notch. Other people successfully perform notching at younger ages but we found it could tear the ear and not make a clean cut. If you are pedigree notching in the USA it must be performed before seven days of age. The piglets don't really object to notching and there is surprisingly little blood created, with the most bleeding seen at the thicker parts of the ear nearer the head than at the ear tip. You have of course now created a wound, so give the notches a spray with an antiseptic and keep an eye out for infection. Clean and sterilise the notchers in disinfectant or boiling water after use.

COMMON INJURIES AND ILLNESS

As with all animals, pigs can suffer from injury or illness. Following good husbandry practice and biosecurity procedures will help keep illness to a minimum and I strongly advise you explore the Internet (www.thepigsite.com) or buy a comprehensive pig veterinary/ health book on the subject. I can thoroughly recommend *Pig Health* (5M Publishing) – my pig bible for anyone thinking of rearing large numbers of pigs or breeding. None of the information replaces qualified veterinary advice, however, and is just brief guidance on a few of the more common or avoidable problems.

It's unlikely that, if you sourced fattening weaners from a reputable breeder, you will have many problems in the four to six months they're with you. If you have a healthy pig, good husbandry and effective biosecurity, then bugs have less chance of getting in and taking hold. Ditto for injuries: healthy pigs heal quicker and are less susceptible to secondary infection. Prevention is the key. Get it right and your first aid kit may remain in the cupboard . . . until you decide to start breeding, anyway!

It goes without saying that if you think what the pig is suffering from is outside your comfort zone for treating and/or you don't have the necessary medicines, then you must call your veterinary surgeon.

TIP

Getting to know your pigs well is essential. You may be able to intervene before an illness becomes critical, and dealing with an injury or illness is much easier if they're friendly.

It is illegal for your friend to come along, diagnose and provide treatment that the vet has prescribed for use on their own pigs. Number one in all situations is to keep an eye on the pig's fluid intake. Always leave fresh clean water near to the pig, even dribbling it into the pig's mouth with a syringe if you have to, as pigs are highly susceptible to water deprivation.

Injuries

Pigs do sometimes injure each other, especially mothers standing on their piglets. The most common cause of a wound with young pigs is play-fighting, and if mixing pigs from different sources, then real fighting. Ears and tails are common injury sites although any-where teeth can grab can succumb. Clean the wound of dirt from the inside to outside so you can see the extent of any damage. Most pigs, if they're eating and know you well, will allow you to do this. Once clean, apply an antiseptic spray and keep an eye on it. If the wound has already become infected, use a prescribed antibiotic spray. Large or deep wounds, e.g. from a sticking-out nail on the fence, may require professional vet-erinary attention. Treat as you would a smaller wound, as sometimes a little blood goes a long way – but if you find the cut is very deep or an abscess has formed, then further antibiotics may be required. Pig wounds should heal remarkably quickly once treated.

Lameness

Lameness can be caused by both injury and illness but the most likely reason is a fall or slip, being trodden on by a heavier pig, bruising of the foot pad or pulling a muscle in thick mud. Remain vigilant for signs of diseases, e.g. Foot and Mouth, especially if you have recently introduced a new pig and/or it's more than one pig that is affected. With mild lameness, i.e. the pig is still mobile, watch frequently to see if it gets better or worse. It's worth washing the leg off, as often it's caused by a stone caught in the trotter or damaged hoof that requires trimming, and if it's not, you can now see any cuts, swell-ings or abscesses on the leg clearly. If the lameness is severe or affecting numerous pigs in the group, then immediate isolation and calling the vet is required.

Coughing and/or Sneezing

Coughing or sneezing may be due to a variety of reasons, not all of them sinister – exactly the same as in humans. Both form part of the body's first line of defence and are normal physiological responses. In a group of pigs, it may be that one has a mild cough and then recovers and all is well, but then another in the group may cough and then recover and so on, giving the impression of a permanent cough in the group. In reality, they're all individually recovering and developing immunity to the bug and all is actually OK, so the pattern of coughing is important in determining the cause and any response to it.

A sudden outbreak of coughing affecting a group of pigs indicates an environmental insult such as dusty/mouldy bedding or an infectious bug, whereas a low-level constant coughing indicates a longstanding, untreated problem, e.g. lungworm. A single coughing or sneezing episode in one pig is probably due to something stuck in its throat or snout. Pathogenic causes of coughing can include bacteria, mycoplasmas, parasites and viruses, and other causes can be immunological, poisoning, fungal or environmental, including heat stress and dust. So administer a dewormer if required, especially if it was not done recently or at the time of purchase. Identify which individual pigs are coughing and note down if they recover. Check your bedding isn't dusty or damp, clean out the housing, disinfect the house and replace bedding. Scrub out water containers to eliminate some of the common causes. You could also try some fresh pineapple, which is rumoured to have anti-viral properties. If the symptoms persist or get worse, call your vet for advice.

Diarrhoea/Scouring

One of the first things you are likely to come across, especially when buying in fattening weaners, is scouring (diarrhoea). Piglets can scour for many reasons – bacterial, viral, parasitological; change of diet; stress – the list is endless. Scouring is often a short-lived occurrence and almost inevitable when you get your new piglets home and put them in a pen full of fresh grass (Figure 5.11). If your pigs are otherwise healthy, only feed proper pig food, keep the water super fresh and wait a few days, Nine times out of ten, the scour will stop and you can relax again. If you do add anything to their diet, then it should be an under-ripe banana to help bind them up. If signs of deterioration in the pigs occur or the scour doesn't start to resolve itself in a few days, call your vet.

Although not highly accurate, you can perform a diarrhoea litmus test to help determine the cause of scouring. Litmus paper, used for checking pH values, is available from pharmacies and drug stores. It can be dipped in a fresh diarrhoea, and if the colour change shows alkaline, then the cause is likely to be bacterial e.g. *Escherichia coli*. If it's acid, then it's likely to be viral, e.g. rotavirus; if it shows a neutral pH, then look for nutritional or parasitological causes and recent stress factors.

FIGURE 5.11 A scouring piglet.

Constipation

Constipation can occur with feed or environment changes, especially if you thought you could feed endless quantities of acorns. Acorns are toxic, especially to piglets. While the odd few won't hurt, larger quantities cause stomach ache and constipation. The pigs may look hunched up, be reluctant to feed, the coat may look rough, and if they do defecate, they may only produce small, hard amounts. Some medications can cause constipation, so look up side effects on the datasheet that comes with all medication. A wet feed with an over-ripe banana and a splash of vegetable oil mixed in often shifts stubborn bowels, specifically purchased for the pig and not from your kitchen in the UK.

Water Deprivation/Salt Poisoning

Another less common but notable condition is colloquially known as 'staggers', which is the result of water deprivation and can have a mechanical or bacteriological cause. The smaller the pig, the quicker the effects of water deprivation will be seen, but it can and does affect all ages of pigs.

The normal levels of salt in a pig's diet (0.4 to 0.5 %) become toxic in the absence of palatable water. Nervous signs develop with the pig staggering about, looking drunk, and it may appear blind. It may fit and convulse if water remains unavailable, with a classic nose-twitching occurring before each fit. Meningitis and death will quickly follow.

Once clinical symptoms are established, the prognosis isn't usually favourable, and careful rehydration of the pig is required if it is to survive. This must be done slowly, as rapid changes in salt levels in the brain caused by sudden rehydration may kill, so small amounts frequently offered is the best way forward. If oral rehydration isn't working out, hydration via the rectum can be performed using a 0.5 % saline solution or warmed water slowly dripped in via a hose inserted reasonably deeply into the rectum. If you are not confident in performing this oddly named 'flutter valve' technique, call your veterinary surgeon or a more experienced person.

Heatstroke can cause similar 'drunk' symptoms, but it is often accompanied with vomiting. Here the treatment is to cool the pig by ensuring access to decent shade, fresh drinking water and spraying it with warm or tepid water – not cold. Too cold and the temperature receptors in the skin assume it's cold weather and try to conserve heat within the inner core, increasing the pig's temperature. A bucket of cold water thrown over a pig with heatstroke is a quick route to death.

If these symptoms appear and you feed lots of fruit to your pigs, just check that it isn't fermenting fruit in their system; it may just be they appear to be drunk because they are! As with humans, a good sleep is then all that is required.

A fourth bacterial cause is associated with recently weaned pigs and discussed in Chapter Nine.

Quick Checklist to Identify Possible Causes of Illness

Think of all possible explanations as to why your pig is unwell and then narrow them down to the most likely reason.

TABLE 5.4 *Some symptoms, possible questions, and a few answers.*

SYMPTOM	SOME QUESTIONS TO ASK	POSSIBLE ANSWERS
Not eating Not drinking Weight loss	Have they stuffed themselves with fresh grass?	They might just be full!
	Have you fed anything different?	They could have tummy ache or just be cautious because they're not used to it
	Is feed mouldy?	Never feed anything you wouldn't put in your own mouth
	Do vermin have access to stored feed?	Rats and squirrels spread diseases such as salmonellosis and leptospirosis
	Have they eaten something poisonous?	Check their pen for strange plants and clay pigeons and be careful where you poison rats
	Is the water contaminated?	Bird faeces, pig faeces and algae might be putting them off drinking
	Is the dung frequency and consistency normal?	Look for other signs of illness, associated husbandry factors, and temperatures over 39°C
	Have they been dewormed?	Parasite burdens lead to weight loss
Lethargic/refuse to leave their bed	Are they lame?	Is it an injury and can you see anything in their pen that may have caused it, e.g. very thick mud/stones?
	What do they look like when you get them up?	Hunched backs, staring coats, shivering, depression, sitting like a dog, arguing with you are signs of illness and pain
Uncharacteristic behaviour	Is their water fresh and accessible?	Water deprivation can make pigs aggressive, so check source
	Are they being bullied?	Make sure all pigs get a fair share of the food
	New pigs in the group?	The dynamics change for a while
Coughing/sneezing Noisy breathing Discharge from eyes or snout	Have you recently introduced new pigs?	The biggest cause of pigs becoming ill is a new pig
	Have they been dewormed recently?	Deworm them if they haven't
	Is their bedding or feed mouldy, damp or dusty?	Check. Disinfect and change if required

SYMPTOM	SOME QUESTIONS TO ASK	POSSIBLE ANSWERS
Scratching and rubbing constantly	Lice and eggs can be seen on the pig hair and skin	Treat twice 10 to 14 days apart with injectable Ivermectin or specific topical spot-on
	Mange mites are most noticeable inside the ears, showing as thick dark wax	
	Ringworm is a fungal skin infection that can be itchy, the most noticeable symptom being ring shapes on the skin	Treat with an anti-fungal, e.g. Clotrimazole, and keep an eye on yourself as it can be passed to humans
Swellings	Is the swelling painful?	It could be an abscess, which can be lanced cleaned and treated as a wound, or a haematoma (blood blister), which can be left alone
Cuts/wounds	Check housing and fencing	Remove any likely cause of injury. Clean wounds and treat appropriately
Vomiting	Recent weather conditions?	Possible sunstroke
	Have they eaten something disagreeable?	Mouldy food, clay pigeon, dead poisoned rats
Colour of urine dark or lack of urination	Are they lame?	Are they holding onto their urine so they don't have to walk?
	Lack of water?	Highly dangerous situation and can cause death remarkably quickly
	Urinary tract infection?	Check for blood when urinating

Answering some or all of these questions will help you identify any possible causes and therefore initiate appropriate treatments, or help you decide to call the vet. Only you know what is normal for your pig, so make use of feeding times and quiet times to get to know your pig intimately. Give your vet as much information as possible to help them make an accurate clinical diagnosis. If you ever become an established breeder, then you will most likely develop a good relationship with your vet and you can administer your own drugs. You can look up the symptoms online, e.g. www.thepigsite.com, or in a veterinary book and treat appropriately, and/or call the vet. Be warned: often the worst-case scenario is displayed on the Internet, so don't panic, and reread the advice given with this in mind.

Isolating the Sick or Injured Pig

Be careful when choosing to isolate your pig. I am always reluctant to isolate the pig completely unless I have no choice. Isolated pigs may become depressed and take longer to get better, so, if at all possible, be creative with your isolation. If it was an injury, use electric fence to separate off another area, to enable the pig to see others; if it was an infectious bug, try and isolate with another from the group or even the whole group and attend to these animals last, disinfecting your boots as you leave.

We had a sow that sustained quite a severe injury; we knew it wasn't infectious, so isolation would have been to benefit her, rather than the whole herd. However, during a previous isolation before farrowing, she suffered from a major case of depression and was a very subdued, unhappy pig, so we had to make a judgment call. We decided to leave her in the pen with one companion and take water and feed to her next to the ark. We had to stand and defend her feed from the other pig, morning and evening, but she didn't get depressed, and when she farrows now, we put her in a pen next to other pigs and this seems to have stopped further occurrences.

CASUALTY PIGS AND DEAD STOCK

One day you may come across the problem of a dead pig, or the need to have a pig quickly and humanely destroyed – sooner if you breed.

In the UK, it is against the law to bury dead stock of any age and you can't throw them out in the household waste. There are a number of options available, but the most common way of dealing with the problem is to call the local knacker man or fell man with approved incineration facilities, and if the pig is still small, some let you take it there yourself in a sealed, leak-proof container or they will collect from the farm very quickly. You could call the local hunt, who may be interested if no medicines have been administered, although it is usually cattle that they want; if you have a large number of animals, you can join the National Fallen Stock Scheme (0845 054 8888) in the UK. Until the animal is collected or disposed of, it must be removed from the other pigs to prevent them from eating it, and covered to stop access by other animals, e.g. cats, foxes, dogs, etc.

For advice on what to do and when with a casualty pig, it is worth getting a copy of the Pig Veterinary Societies' booklet, *The Casualty Pig* (ISSN0956-0939) to keep on your bookshelf for reference when required.

In the USA, there are five mortality management methods that are practised. However, each state has its own guidance rules on each method. The five methods consist of landfill at an official landfill site; on-farm burial that is not near to any surface waters; rendering to create meat or bone meal as a feed ingredient; incineration, which can be on-farm with a specially designed burner; or composting in biosecure carbon piles. Some of these practices involve a fee for carcass collection and/or processing.

SOME LESS COMMON BUT NOTABLE DISEASES

There are some less common diseases that are worthy of knowing about. In the UK, some of them have now been eradicated or are currently not present, and also in some states within the USA. In the USA there may be compulsory vaccination programmes in place, so when you first get pigs, consult your State Veterinary Surgeon. These diseases are less common in extensive rearing systems, but they can and do still occur, so you cannot be complacent. A decent pig veterinary manual will cover each disease, providing additional, more detailed information.

Porcine Epidemic Diarrhoea (UK and USA)

Porcine epidemic diarrhoea is caused by a coronavirus named PEDv. Two different types are recognised: PEDv Type I only affects growing pigs, whereas PED Type II affects all ages including sucking pigs and mature sows. Type I is seen as a low-level endemic disease in the UK with pro-active surveillance only detecting very low numbers annually. Type II is a more virulent strain that caused the death of over 1 million pigs in the USA in 2013–14, with up to 100 % mortality seen in piglets less than seven days old.

PEDv damages the villi in the pig's gut, thus reducing the absorptive surface, resulting in a loss of fluid, diarrhoea and dehydration. Up to 100 % of sows in a herd may be affected, showing mild to watery diarrhoea, after the PEDv enters a herd, but a strong immunity develops over two to three weeks and the colostrum then protects the piglets. The virus usually disappears spontaneously from small intensive breeding herds; however, in larger outdoor breeding herds not all the pigs may become infected first time round and there may be recrudescence.

Transmission

PEDv is transmitted via the faecal–oral route and infected pigs shed enormous amounts of the virus for seven to nine days. Transmission may be through direct contact with infected pigs or indirectly by exposure to infected pigs' faeces, which may persist in cool, damp organic matter for up to a month. The virus is killed by most common disinfectants such as Virkon S™. It is not infectious to humans.

Symptoms

The clinical signs of disease are very age specific and much more severe in younger animals. In very young piglets there is profuse, watery diarrhoea, without blood or mucus, which is often yellow in colour, often accompanied with vomiting or anorexia, which may lead to death in up to 100 % of the piglets less than a week old. Pigs over a week of age typically recover. When older animals (nursery, grower, finisher, sows, boars) become infected, they may go off feed for two to four days, have loose faeces or watery diarrhoea and vomit – with a death rate of 1 to 3 % in the post-weaned animals. The incubation period is typically one to four days and the first clinical signs in a herd are seen approximately four days after PEDv enters. There are other diseases that cause very similar clinical signs, such as coccidiosis, transmissible gastroenteritis, rotaviral diarrhoea, *Clostridium perfringens*, enterotoxemia and *E. coli* scours. It is essential to submit proper samples to a veterinary diagnostic laboratory for diagnosis if you have a diarrhoeal problem on this scale.

NOTE: High mortality in piglets less than seven days old would be an early signal of virulent PEDv Type II currently not seen in the UK.

Diagnosis

This is based on the history, clinical signs in the different age groups and examination of faecal samples/dead piglets for evidence of PEDv Type I and Type II by laboratory diagnosis.

Treatment

No specific treatment for PEDv is available. Affected pigs should be kept warm, dry and well hydrated with oral electrolyte supplementation. In very young animals, treatment is usually futile. If secondary bacteria complicate the clinical disease, or are likely to, then broad-spectrum antibiotics may be prescribed. If the virus enters the herd for the first time, it is important to ensure that all the adult animals become infected at an early stage to allow an early immunity to develop. This can be achieved by exposing sows to the diarrhoea three times, two days apart via the drinking water. Mix scour or contaminated material into a bucket of water and use this as the source.

Prevention

Strict biosecurity and sanitation are the best means of prevention. A vaccine has been produced in the USA but its efficacy and how best it is used is still being evaluated.

Porcine Reproductive and Respiratory Syndrome (UK and USA)

Porcine Reproductive and Respiratory Syndrome (PRRS) is caused by an arterivirus commonly known as PRRS virus (PRRSv), which was first isolated in 1991. The disease itself was seen before the virus was isolated in the USA and known as 'mystery swine disease' and 'blue ear'. In sows, it causes reproductive failure and in young pigs it causes a severe respiratory tract illness and is estimated to cost the United States swine industry around $644 million annually, and €1.5b across Europe. There are two distinct strains – North American and European – both causing similar symptoms. PRRS infects all types of herd, including high or ordinary health status and both indoor and outdoor units, irrespective of size.

Transmission

The virus is spread by nasal secretions, saliva, colostrum, milk, faeces and urine and studies suggest it can be airborne for up to 3 km (2 miles). A carrier state exists in the pig that can last for two to three months or more in some individuals. Adult animals shed PRRSv for approximately two weeks and growing pigs for up to two months. PRRSv may infect foetuses from mid-pregnancy onwards. Artificial insemination (AI) could be a potential method of spread if semen is used when the virus is present in the blood, particularly during the first three- to four-week period following the breakdown of an AI stud. However, AI studs are very aware of this and test for it.

Symptoms

Subclinical infections are common, with clinical signs occurring sporadically in a herd. Clinical signs include reproductive failure in sows such as abortions and giving birth to stillborn or mummified foetuses, and cyanosis of the ear and vulva. In neonatal pigs, the disease causes respiratory distress, with increased susceptibility to respiratory infections such as Glässer's Disease.

When the virus first enters the breeding herd, the disease is seen in dry sows, lactating sows and nursing piglets. You may see short periods of anorexia in up to 15 % of the sows but lasting 12 hours at a time; elevated temperatures of up to 40°C (105°F); blue colouring and swelling of the ears; infertility shown by increased number of returns to service (coming back into season rather than conceiving or maintaining a pregnancy); and coughing/respiratory symptoms.

Clinical signs in farrowing sows include a reluctance to drink, agalactia (lack of milk production) and mastitis (infection in one or more of the mammary glands/udder), farrowing two or three days early, discoloration of the skin and pressure sores associated with small blisters, lethargy, up to 15 % mummified piglets if infected in last four weeks of pregnancy, up to 30 % increase in stillborn piglets and very weak piglets at birth.

The piglets are born in a very weak condition and rapidly become hypoglycaemic (lacking glucose for energy to function) because they are unable to get to the teat. They may have a sticky brown material over the eyelids and very occasionally small blisters on the skin. Scouring, pneumonia and coughing are commonly observed, but the longer they live, the more likely they are to increase in quality and survive.

Growing pigs post-weaning can become infected and shed PRRSv for up to four weeks, with clinical disease seen from 4 to 12 weeks of age. They are reluctant to eat and show signs of malabsorption with wasting signs. They usually progress through coughing to pneumonia. Mortality can rise to over 12 % and secondary bacterial infections become evident from 12 to 16 weeks and form abscesses on the lungs. These infections may spread to other parts of the body, particularly joints with increased lameness.

Diagnosis

Aujeszky's Disease (AD) when it first enters the herd can be confused with PRRS, however, nervous signs are present with AD but not with PRRS. A serological test will differentiate between the two.

Treatment

All treatment is aimed at preventing secondary bacterial infections with aggressive antibiotic therapies, usually until immunity is built up. Keeping the pigs hydrated is also key to survival and interventions may be required if they are reluctant to drink.

Prevention

There are vaccines that can be used in conjunction with herd-specific eradication and control programmes. This is a job for your veterinary surgeon, and a multi-pronged approach specific for your farm must be devised.

Post-Weaning Multi-Systemic Wasting Syndrome (UK and USA)

Post-weaning Multi-systemic Wasting Syndrome (PMWS) is caused, in part, by a porcine circovirus (PCv). There are two types: Type I causes no known disease and Type II can be found in the lesions that are caused. Most infections are sub-clinical (not seen visually) in herds and it is not known why clinical symptoms are sometimes seen. However, they are more likely to develop lesions if another virus, such as porcine parvo virus (PPV) or porcine reproductive and respiratory syndrome virus (PRRSv) is also present, but most pigs infected with either of these viruses don't usually develop PMWS.

Transmission

Transmission is via the faecal–oral route and may be direct transmission from another pig in the group or indirectly transmitted through dirty boots, etc.

Symptoms

Symptoms are only seen in weaners and growers, usually starting around six to eight weeks of age. The piglets lose weight and gradually become emaciated, growing long hair. They may become pale or jaundiced. Diarrhoea, respiratory distress, incoordination and sudden death are not uncommon. Depending upon herd size and number of susceptible piglets being weaned, it may take 6 to 12 months to reach its peak before declining.

Diagnosis

As PCv is present sub-clinically in most herds only, diagnosis is based upon the presence of PCV Type II histological lesions in lung, tonsil, spleen, liver and kidney tissues *post mortem.*

Treatment

Like with all virus infections, palliative care and hydration are your first priority, with supportive antibiotics if a secondary bacterial infection occurs.

Prevention

There is no current vaccine, so maintaining good husbandry and biosecurity between groups is important. The virus is susceptible to disinfection but can remain in the environment for a considerable time.

Brucellosis (USA)

Brucellosis in pigs is caused by the bacteria *Brucella suis*. There are different biotypes, with the most common biotypes that cause disease in pigs being biotype 1 and biotype 3. *Brucella suis* spreads slowly between and within herds and, although a severe disease, you should be able to keep it out of your herd if you take sensible biosecurity precautions. If it does get into your herd, it is difficult to eliminate, causing long-term reproductive losses, and some biotypes (1 and 3 in particular) are zoonotic, causing a serious illness in humans.

Transmission

Transmission occurs through direct or indirect contact with infected bodily secretions such as semen and vaginal discharges.

Symptoms

The earliest sign is usually a few pregnant sows aborting, followed by an increase in the number of sows returning to service, and/or vaginal discharge. If an infected boar has caused infection in a sow, then she will return to service approximately 30 to 50 days post-service. If sows are already pregnant, abortions may occur at any stage and the environment may be contaminated by the vaginal discharges. The sow's infected reproductive tract will eventually clear spontaneously. Boars' testicles may swell and they may shed the bacteria in their semen for prolonged periods of time, causing permanent, irreversible damage. Some growing pigs and adults and some nursery piglets will start to develop partial or total paralysis in their hind quarters, from the infection causing damage to the spine. Some pigs may become lame with swollen joints.

Diagnosis

Clinical disease history from all age ranges in the herd is commonly used to identify an infection, but laboratory tests have to be undertaken to make a definitive diagnosis.

Treatment

Treatment with antibiotics is not very effective and the best course of action is to cull the affected pigs. Even culling all pigs should be considered, and in some countries the decision would be made for you.

Prevention

Due to the low incidence of disease, a vaccination in pigs hasn't been developed. There are national or regional control and eradication programmes in place in a number of countries, including the USA, to eventually eliminate this disease by the compulsory slaughter of herds in which the organism is diagnosed. This disease has been eradicated from the UK and Ireland.

Aujeszky's (USA)

Aujeszky's Disease (AD), or pseudorabies (PR), is primarily an infection of pigs, and pigs are its only known reservoir host. It can be transmitted naturally between pigs, cattle, horses, dogs and cats – all of which develop nervous signs and rapidly die. The UK and Canada are free of AD but cannot be complacent as it would only take one import from infected EU countries or states within the USA to become reinfected.

Transmission

The primary method of spread between pigs is aerosol and nose-to-nose contact. It is not spread in faeces or urine. The causal agent is a herpes virus and like the herpes viruses of humans, the AD virus can lie dormant in the pig's nerve cells. Stress can reactivate it and the pig starts to shed virus again.

Symptoms

In the breeding herd, the earliest sign is usually a few pregnant sows aborting. Farm dogs and cats may develop a severe acute nervous disease and then die. Nursing piglets also develop an acute severe fatal nervous disease, usually exhibiting a rough-haired appearance. They stop feeding and within 24 hours develop nervous signs, salivate excessively, convulse and usually emit a high-pitched squeal. Slightly older piglets may walk in circles, sit like a dog or paddle their legs while lying on their side. Some may vomit and/or have diarrhoea and then die. Mortality in piglets may be as high as 100 %. Fully weaned pigs may additionally exhibit respiratory illness and be more susceptible to secondary bacterial infections including mycoplasmas. They may also have a high fever up to 42°C (107°F), but only approximately 10 % will die and most recover in five to ten days.

In grower and finisher pigs, the earliest signs are depression, lack of appetite, staring coat and fever of 41 to 42°C (106 to 107°F), with respiratory signs similar to weaned pigs. Most (98 %) recover in five to six days and start to grow well again.

Diagnosis

Clinical signs are used to make a presumptive diagnosis, especially those seen in new-born piglets and if any dogs, cats, cattle or horses are affected. After that a laboratory diagnosis is advisable, which may be compulsory in some countries.

Treatment

There is no treatment available specifically against the virus; however, antibiotics will prevent secondary infections particularly of the respiratory system in the grower and finisher pigs.

Prevention

There are vaccines available, although their use in countries free of Aujeszky's may be prohibited by law.

Classical Swine Fever (UK)

Classical Swine Fever (CSF), also known as hog cholera (HC), is a specific viral disease of pigs and is a notifiable disease in most countries. CSF is caused by a pestivirus and is highly contagious, quickly spreading through a herd. The USA is considered free of CSF and the last case in the UK was 2000 at the time of writing, but it is endemic across certain parts of Europe, and with the free trade agreements, the UK is at a continual risk of reinfection.

Transmission

Transmission is via direct or indirect contact with infected faeces or bodily fluids including sows to piglets and via ingesting undercooked or cured infected meat/meat products.

Symptoms

In an unvaccinated herd, almost all the pigs of all ages are affected. It causes a generalised disease of fever, malaise, lack of appetite, diarrhoea, paralysis, abortion, mummification and the birth of piglets with tremors. A consistent early sign across the herd, which persists until death, is a high fever, over 42°C (107°F). Mortality is usually high.

Diagnosis

Clinical signs in the first instance and laboratory diagnosis to differentiate from other diseases with similar symptoms.

Treatment

No specific treatments are available.

Prevention

There are vaccines available but their use in countries free of CSF is usually prohibited so that serology can be used to assist in a diagnosis in the lab if required.

Prevention is via strict biosecurity and sticking to the swill ban in the UK. Control is government led via national eradication programmes and usually the immediate culling of the pigs on the affected farm.

Glässer's Disease (UK and USA)

Glässer's Disease is caused by the bacterium *Haemophilus parasuis*, of which there are many different subtypes. It is ubiquitous and found throughout the world and sub-clinically present even in high-health herds. Most herds have a level of immunity, with sows passing a strong maternal immunity to their piglets which lasts until they are up to 12 weeks old. Piglets usually become sub-clinically infected while still under immune protection from the sow. Outbreaks can occur in nursery piglets if the natural immunity wanes before they become infected, often in litters born to gilts. They can develop a severe disease or it can be a secondary infection to another disease that has made the pig weak.

Transmission

Transmission is via respiratory secretions.

Symptoms

It is rarely seen in sows but may be seen in gilts, causing slight swellings over the joints and tendons, resulting in a stiffness or lameness.

The acute disease in piglets causes a rapid depression, elevated temperature, reluctance to eat and get up, two to three short coughing episodes and possible death. In chronic disease, the piglets are pale, hairy and have a poor growth rate, arthritis symptoms of varying degree and sudden death remains a possibility. In older growing pigs, nervous signs may additionally develop, including convulsions, heart infections, peritonitis and meningitis.

Diagnosis

In the living pig this is difficult, because it is similar to *Mycoplasma hyosynoviae*, an infection of joints and tendon sheaths. Diagnosis *post mortem* in the laboratory will differentiate between the two conditions.

Treatment

Because this disease can be very difficult to differentiate from a mycoplasma infection, a combination of antibiotic treatments is often prescribed to cover both infections.

When the disease has become a problem in the herd, the veterinary surgeon may pre-scribe antibiotics at point of farrowing, after farrowing or in the piglets' creep feed. Autogenously made vaccines specific to the herd may be produced and given to the sow in chronic situation.

Foot and Mouth Disease

Foot and Mouth Disease (FMD) is the most important restraint to international trade in animals and animal products and is taken ruthlessly seriously. FMD is so important because it is highly infectious, spreads rapidly throughout cloven-hooved animal popula-tions and over long distances on the wind, and hence it is difficult and costly to control. Foot and Mouth Disease virus (FMDv) is a picornavirus that causes virus-filled blisters or vesicles to form in the mouth and on the feet of susceptible animals.

Transmission

Transmission is via direct or indirect contact with infected animals, including via aerosol, and may be carried for miles in the wind. The virus is present in great quantity in the fluid from the blisters, and it can also occur in the saliva, exhaled air, milk and faeces.

Symptoms

The pre-patent period, which is the time between infection and the appearance of symp-toms, varies between 24 hours and over 10 days, with an average time of three to six days. Pigs will often remain lying down and be reluctant to move and will squeal loudly in protest if forced to do so. Blisters form on the upper edge of the hoof, where the skin and horn meet, and on the heels and within the cleft. They may additionally develop on the snout and the tongue.

Diagnosis

It is impossible to differentiate clinically from other vesicle-forming diseases, and if sus-pected, then a licensed government vet or the state vet must be contacted, who will send the appropriate samples to one of only three laboratories in the world capable of making a confirmatory diagnosis.

There is no treatment and all cloven-hooved animals on the farm must be culled. This is performed under strict government supervision. Restocking is not allowed for a minimum of six weeks.

Prevention

In countries where the disease is endemic, and in high-risk areas, a vaccine may be used to protect the breeding stock. In the USA and UK prevention is only via the governments' robust emergency control plans.

HUMAN HEALTH

Seems an odd topic to include in a book on pig keeping, but there are some health implications of which you need to be aware. There are diseases or organisms that can be transferred from pigs to man, known as zoonotic diseases. Governments spend a lot of money each year researching into ways of minimising the transmission of these into the food chain. Most farms supplying meat to the general public have to sign up to health schemes to monitor their farms once they have over a certain number of pigs, and if these organisms are found on the premises, they have to jump through varying hoops to be able to carry on trading.

The transmission routes – i.e. how the bug spreads from the pig to you – can vary, so some may be primarily foodborne, some spread via direct contact with the pig and their faeces and urine, and some require prolonged close contact or breathing in the same air. Just to complicate the picture, some cause clinical disease in the pig and are therefore more obvious, while some do not affect the pig at all and so it is perhaps harder to remember they may be present.

The most vulnerable to the risks of catching disease are the very young, the very old and the immune compromised, e.g. people on anti-rejection drugs after a transplant. Don't deny children the experience of interacting with the pigs, but be a bit more careful about their hygiene. Children who live on farms will have more immune resistance, as they are constantly being exposed to different bugs, but be super careful with visiting children.

There have been a few high-profile cases of children's petting farms, butchers' shops and restaurants being temporarily closed because of human disease outbreaks. Often they involve a bacteria called *E. coli* 0157 which ruminant livestock carry but do not suffer disease from. However, in the young/old and immune compromised, this can kill. So while these cases are rare, they can and do happen and you need to be aware of them. The Health and Safety Executive (HSE) has published some very good guidelines on

TABLE 5.5 Some zoonotic organisms harboured by pigs that you need to be aware of.

ORGANISM	SYMPTOMS IN MAN	TRANSMISSION	PREVENTION	COMMENTS
PARASITES				
Ascaris suum	Coughing, vomiting or diarrhoea	Ingestion	Hand washing	Eggs/oocysts can survive for years in soils
Cryptosporidium parvum	Diarrhoea and/or vomiting		Wash, peel or cook vegetables	
Toxoplasma gondi	Muscles aching, 'flu-like symptoms		Do not eat raw pork	Serious complications for pregnant women
Trichinella spiralis (not known to be in UK)	Muscles aching, cysts in brain			Do not feed pork products to pigs
BACTERIA				
Brucella spp. (not in UK)	'Flu-like symptoms	Handling of secretions from aborted, or newborn piglets	Hand washing	Primary risk from outdoor and feral pigs
Erysipelothrix rhusiopathiae	Raised dark lesions on hands	Direct contact		Possible spread by birds
Staphylococcus aureus including MRSA	Lesions on skin			Special handwashes are available in chemists and drug stores
Streptococcus suis	Fever, lesions on skin			
Leptosporidia spp.	'Flu-like symptoms	Direct contact with urine		Use PPE for confirmed cases
Campylobacter jejuni	Diarrhoea, fever, vomiting	Ingestion	Hand washing, cook meat thoroughly	
Salmonella spp.				
Yersinia enerocolitica		Ingestion or direct contact		
VIRUSES				
Influenza	Coughing, fever, aching, chills	Inhalation	Hand washing, consider face mask	Can be passed back and forth

personal safety for the public on farms, which is worth having a look at. Type www.hse.gov.uk/pubns/ais23.pdf into your browser.

What you need to know is that in the majority of cases good personal hygiene, cooking the meat properly and effective on-farm biosecurity keeps the risk of transmission extremely low. As long as you are aware of the risk, there is no requirement to attend your pigs in a biohazard suit – just wash your hands after handling the pigs and minimise hand-to-mouth contact until you do. Don't rely on alcohol-based wipes and sprays, as they have no effect on some of the more serious bugs and you are effectively killing off the less harmful ones, leaving those that can cause harm to do their worst. Use soap and running water every time.

BASIC PIG-HANDLING TECHNIQUES

TIP

Try and move pigs only when you have time. Plan ahead. If you need to swap over pigs for the boar to service, for instance, choose a day you are not working and stick to it. Don't be tempted to 'quickly' do it before you have to go to work.

What is most likely to stop kind handling is frustration. Pigs can be highly frustrating, particularly our Gloucestershire Old Spots for being too slow and the Mangalitza for being too fast! Always give yourself plenty of time when moving your pigs from A to B – if you are under any time pressure at all, then you are more likely to lose your temper and they are then less likely to cooperate. Of course if a pig requires isolating and you have to move it, then you will have to kick a tree or something if you get frustrated.

COGNITIVE PSYCHOLOGY

Pig psychology, pig behaviours and the whole pig approach is in practice an advanced stockman's eye. Pigs are constantly sending out signals of their mental health, their emotional state, their physical health, and these can be read by humans and interpreted. You notice a behaviour/symptom; decide what your reaction should be; perform that reaction. The *Commuter Pig Keeper* can get easily distracted with routine chores (especially before work). You may just not notice something or notice but do nothing about it, which equals the same outcome. On a simplistic level, for example, you want to get a pig pregnant, you notice she is in season on a Tuesday morning but do nothing about it until the weekend. By then it's too late and you have missed your chance for another three weeks. OK, that is annoying, but it is not life threatening, what if it was a bout of diarrhoea or lameness?

The commercial pig sector uses the body language of pigs as an indicator of good and bad health, although they tend to promote the 'hands-off' approach to minimise stress – which is sensible when you have thousands of pigs. They use the instincts of the pig to their advantage, such as moving pigs in familiar social groups from A to B, through a series of passageways and gates, with a person with a large movement board the width of the passageway driving the pigs from behind. They make sure that nothing unfamiliar is placed *en route* causing a backwards movement of the pigs, and use a familiar bedding to disguise areas on the floor such as drains, lorry ramps, etc. that would spook the pigs. To do this you have to know what makes the pigs anxious and is likely to cause stress. These natural instincts are rightly exploited when it makes the farmer's life easier and so less stressful for the pigs, but, unfortunately, they are actively avoided when it suits the farmer, increasing the pigs' stress. Many sows are denied freedom when farrowing, which suits the farmer. Allowing the sow to build a nest means she has to be able to move around, which takes up precious space, and means there is more bedding to dispose of; handling piglets for routine procedures such as iron injections, teeth clipping and tail docking means having to increase personal protection.

The smallholder has the privilege of using the 'hands-on' approach to minimise stress. With most breeds, the more the sow knows you, the less likely she is to have a problem with you handling the piglets or loading her in a trailer for the first time. Even breeds that are fiercely protective, such as the Duroc or Mangalitza, should trust you enough to take them to a secure pen for a while during any piglet-handling procedures.

Advantages to understanding pig behaviour:

- You can talk to your pigs and they can talk to you. It is much easier to find a solution to a situation if you can communicate.
- Promotes trust from the pig – if you are working with their instincts, the whole situation becomes easy without anyone being fraught.
- Being able to understand pig language will assist you in outwitting them when you wish them to comply, e.g. loading in a trailer.

Pigs become an awful lot easier to handle if you know how they think and 'what makes them tick'. They are relatively intelligent and impossibly strong, which can be an awesome combination if you don't handle them correctly. Pigs are sentient beings and have emotions such as pleasure, fear and stress and quite a long memory, so in every procedure you wish them to perform, you *have* to reassure them that it is safe to do. If you were suddenly plucked out of bed and asked to walk a tightrope strung high up across a major road, you would understandably stop at the point you felt unsafe. If, as you slowly went to the rope edge to have a look, someone started shouting and pushing you or poking you with a stick, perhaps hurting you in the process, then you would understandably back up to the previous point you felt safe. If at a later date you found yourself in the same position, you would remember the pushing, pain and sense of fear and not even go to the edge and have a look. It is no different with pigs, so try to see things from their perspective. For example, you have contained them for months in an electrified pen with an electrified gate and it has confined them beautifully. You now want to take them out to load in the trailer to get to the abattoir or a show on time, so you open the gate and the pigs won't cross the line where the fence was. Why would they? The last time they went near that area it really hurt. You are now aware of the time and start to get frustrated, and so *your* behaviour changes, alerting the pigs to a possible danger from other sources as well as the 'biting' fence. The pigs rush back to the safety of their ark and for the first time ever you climb in and shoo them out, yet another change in *your* behaviour. You might even shout or tap them with a stick to assist forward movement. This situation is entirely avoidable and entirely your own fault; don't make the first time you do anything with your pigs that requires their co-operation one in which you are under a time pressure.

NOTE: All welfare codes for pigs state that you must never strike or poke with a pointed stick or electric prod. Pigs must be allowed to move at their own pace, using only gentle encouragement.

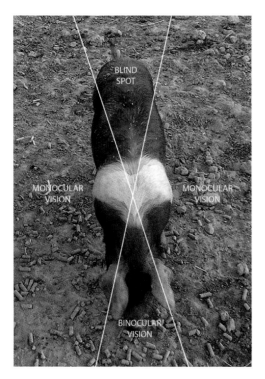

FIGURE 6.1 A pig's monocular and binocular vision.

A Pig's Eyesight

It might be useful to understand how a pig's eyesight could affect its response to a situation. A pig's eyesight is said to be atrocious – but why?

Having eyes either side of their heads means they have the ability to see in monocular (using each eye independently) and binocular (using eyes together) vision. This increases their panoramic vision to more than 300°, which is an excellent survival tool for finding food and detecting potential danger and minimises the size of the blind spot (Figure 6.1). However, this ability leads to a decrease in the amount of binocular vision, which is used to see directly in front of them and gives them the facility to perceive vision depth. It is thought that pigs can view in colour as they have both rods and cones within the eye, but what colours and shades of colours they can detect isn't known. The ability to see in full colour probably isn't necessary as they have a powerful sense of smell and garner enough additional information about their surroundings by rooting and sniffing. If something in the pigs' immediate environment has changed, e.g. a pile of food, then the snout will know. It's therefore important as a handler that you are aware of the sights beside and the smells in front of a pig to assist you in viewing how the pig itself ascertains its position of current safety. So a pig's eyesight isn't atrocious for its own survival needs, but would be for our own.

BASIC HANDLING

Handling Piglets

First of all, if the piglets are still with their mother, then, in most cases, the mother needs to be removed to a safe distance. To minimise stress to the mother it is better that she cannot hear them either, although this may be difficult as a piglet's squeal can reach 120 Db. Very young piglets can be handled by gentle physical restraint, and depending upon whether they are used to being touched and lifted up, they can be picked up around the barrel of the body, supporting their underbelly and held in your arms, as you would a small dog. Pigs do not carry their young like a cat or dog, so to a piglet, being lifted off the ground instinctively means a predator has picked them up and their only protection is to alert the mother by screaming. For this reason, unless they are used to being lifted and have learned that nothing horrific happens, they will scream loudly, making ear defenders a wise purchase.

An alternative method that is useful for piglets between two and eight weeks and often quietens the piglets down so they don't scream, is to hold them upside down by *both* back legs (Figure 6.2); it seems to make them docile for reasons I do not know. You catch them by grabbing one back leg and then getting hold of the other back leg before the majority of the pig's weight leaves the floor. You must hold them just above the hock region: not only is this a natural handle shape that prevents slipping if they struggle, it also protects the ligaments in and around the hock region from bearing too much weight. Always hold them with their spine towards your body so that if they do fancy giving you a quick nip it won't be on your private parts! Don't carry them for any length of distance like this though; it is a quick trailer-to-gate method or to quickly check underlines, etc. Personally I would only practise this technique up to eight weeks of age, 12 weeks if a smaller Kunekune-sized breed perhaps; after that you risk shearing ligaments and causing a hernia.

FIGURE 6.2 *How to safely pick up a young pig by its hind legs.*

Older Pigs

Once over eight to 12 weeks of age pigs are a bit heavy to lift and hold away from your body, plus it would be too stressful on their hind limbs. Forcing a pig to do something by sheer strength is impossible and will add considerably to the stress of the pig by even trying. The simplest method to move an older piglet from A to B, which works most (ish) of the time, is to coerce your pig with food, especially sweet treats like soft fruits or apples. As long as you are not asking them to perform anything too stressful, such as load into an unfamiliar trailer, then this works well in an enclosed, safe environment for activities such as crossing previously electrified fence lines and/or moving to a new pen. It can really help to make the gateway look different by putting some of their bedding over the area or kicking the mud about to make it look different and then putting the food in a line. If you are clever and wait until they are hungry, then before you know it they will have crossed the gateway.

Always end the experience on a good note and feed something yummy before releasing.

LOADING INTO A TRAILER

Preparation, preparation, preparation! The end objective to loading is to have your pig(s) inside the trailer with as little stress to both the pigs and you as possible. If you get stressed, so will they. I appreciate you may not have a trailer to practise loading prior to going to slaughter, but you can practise getting them out of their pens, which is often the part they are the most hesitant to do, and walking them to the point the trailer will be located. Give them a treat and take them back. Often you will identify a 'lead pig' and the others will follow it, so concentrate on making the lead pig be where you want it to be. Prepare a route to contain straight lines and curves where possible, as pigs like to see where they are going, and practise it a few times. If you prepare the way approximately one hurdle wide, so much the better, as when you take the pigs along the now-familiar route, they will stop still when they see the unfamiliar trailer. If you are following behind with a hurdle, you can now just stand on it and let them take their time to investigate. When the trailer is *in situ* it should contain some familiar bedding inside and also have some on the ramp; our pigs are used to straw, so we use straw. It should also be hitched up if it doesn't have four wheels or it will tip down as they go on the ramp, scaring them, and any means of escape on either side should be blocked. If the pigs are well handled and trust you, they should not 'freak out' but just slowly investigate. Of course you cannot take all day to move them, so slowly walk behind them with the hurdle. If they stress, then stop for a minute, let them settle and then push on. Eventually, you should have the pigs on the ramp and the hurdle blocking the exit; at this point they should load, but still let them take their time. If you start suddenly rushing to put them in, then they may get scared and try to go back through you. When they do go inside the trailer, quickly but quietly shut the trailer gates and the ramp.

Using a Triangle of Hurdles to Load

If you can, get them inside a triangle of hurdles within their pen; you may have to tie the hurdles together for ease (carry a sharp penknife for the other end). Then move the hurdles like a rolling road block until you reach the straw-covered trailer ramp; it is tiring to perform over long distances, so get the trailer as close as possible. If the gateway isn't big enough for the triangulated hurdles to go through in one piece, then open one side to get them through the gap and then shut them up again into a triangle once you are through. You may have to stand on the hurdles intermittently en route to stop the pigs lifting them and going underneath but you can stop when required to let the pigs calm down if stressed without them getting away. Once at the trailer, allow the pigs to look at it for a minute or two. Put a little treat on the floor to distract them but only a tiny amount if the destination is the abattoir, and immediately open the triangle of hurdles so that the hurdles are now in a [ ] shape and the trailer and trailer gates are at the open exit. Slowly move the [ ]-shaped hurdle arrangement to either side of the trailer ramp, reducing the amount of space the pigs have. Throw a couple more treats onto the ramp to encourage movement.

Common Sense

Never strike your pigs. It will not get you anywhere and they will be even harder to load, as they will associate trailers with a beating, or if this is their last journey, it will add to their stress and possibly affect the taste of the meat. The calmer you are, the more likely they are to load without stress. They will eventually go in. Don't contemplate your navel when they do, quickly and quietly close any internal gates and the gates prior to the ramp, pat your pig through the jockey door or ventilation panel and give another slice of apple. Positive reinforcement!

Most pigs, once they have loaded regularly and always enjoyed a smooth ride, will enter a trailer quietly and easily with simple stick and boarding and a couple of slices of apple. The only time you cannot use food to entice is if you are loading on the morning of slaughter, when time, patience and encouraging pats are your only choices and the tiniest pieces of apple perhaps.

Wheelbarrowing and Tail Lifting

You may have seen people lifting the rear legs of medium-sized pigs when they will not load into a trailer or go into a pen, and pushing them 'wheelbarrow-style' to the destination. This is not something we do, as we have the correct set-up and so don't have to. We have, however, had to do it reluctantly a couple of times with other people's pigs when they needed help loading and just wouldn't go in any other way. I have not seen or read enough evidence that it is safe for either the pig or its handler to do so. The weight of a middle-sized pig walking on just its front legs, and the weight of the back bearing down on unsupported hips, seems not quite right. When I asked one old pig farmer about the practice he just said, 'It works, dunnit.' Again we don't like the idea of lifting the tail and pushing on it, as this works on the principle of pain avoidance by the pig, which is not how we wish to do things and certainly adds to the pig's stress.

STICK AND BOARDING

Also known as bat and boarding. The idea is to move your pigs from A to B with the use of a board to give direction and a stick with a curve at one end as gentle acceleration of forward movement. When not moving in a straight line, try to move a pig in a clockwise direction, holding the board in your left hand and the stick in your right hand. Stand on the left of the pig next to its middle (Figure 6.3).

FIGURE 6.3 A pig course delegate being taught how to position herself for stick and boarding.

Tap the pig gently with the stick on its ham or just behind the shoulder to ask it to move forward, and as it starts to walk use the board to steer it the way you want to go. When the pig is moving in the correct direction keep the board to the left of its head and tap with the stick gently if required. Do not rap its hocks and genitals or hurt it if you don't get an instant response. We have seen too many 'experts' do this and it is unacceptable: the whole experience should be a pleasant one. If the pig tries to turn around or go the wrong way, use the board to block its field of view, and when all is well again with the direction put the board back to the left-hand side. You will sometimes get left behind and find yourself looking at the rear of your pig to a greater or lesser degree. Never panic, just assume the correct position as soon as you can without spooking the pig, especially the young ones. As with all these things, practice makes perfect (ish) here, so take the time to move your pigs around the farm with the stick and board instead of food. Gain confidence by refining your skills on slower-moving, older pigs if you can, before shattering it again on the young speedy ones.

Start off with short sessions, as you have no wish to frustrate your pig. An ideal time is when you need to move it, and rather than reach for the food first, have a go at using the stick and board. You do not need to buy a board straight away; any flat object that is easy to hold can be used. There is more on how to train pigs using the stick and board for the show ring in Chapter Fourteen.

PEDIGREE IDENTIFICATION

7

I'll talk more about breeding pigs in Chapter Nine. If you intend to breed, then my advice is to buy registered pedigree pigs. The reason for this is that the pig should already meet the minimum requirements of the breed and has been considered by the breeder to be of suitable quality and a good representation of the breed. It therefore seems sensible to mention how to identify pedigree pigs from their ear markings and the registration process so you don't buy something unsuitably identified.

REGISTRATION AUTHORITIES IN THE UK

The British Pig Association (BPA) controls the registration and management of all pig breeds in the UK, both modern and traditional, with the exception of the British Lop and Kunekune pigs, whose clubs deal with their own registration. The breeds currently recognised by the BPA are the Duroc, British Landrace, Piétrain, Berkshire, British Saddleback, Gloucestershire Old Spots, Hampshire, Large Black, Large White, Middle White, Tamworth, Oxford, Sandy and Black, Welsh and Mangalitza. To become a registered breeder, you will need to join the BPA or British Lop Society or Kunekune Society.

When you join you will be asked for prefix options for when you start naming your pedigree pigs. Ours is 'Tedfold'. You will then be allocated with a three-letter breeder's code called Herd Designation Letters (HDL) – ours is GLN – which is used on your ear tags and tattoos in addition to your legal herd mark number requirement as a part of the pedigree identification. This is also your code to enter the BPA members' area on its website. You can then manage your herd registrations completely online, although you can do it by post or by telephone if you wish.

Specific Breed Clubs in the UK

There are specific breed clubs in the UK, but with the exception of the British Lop and Kunekune breeds, none of them is a registry and all are optional to join. The BPA breeds not listed in Table 7.1 do not have their own specific breed clubs yet.

The UK is in an enviable position of breeding some of the finest pigs anywhere in the world. We also have the largest variety of pedigree native breeds and so it's no coincidence that all of them have their own specific clubs. We also have the British Kunekune Pig Society (BKKPS), with probably the largest membership of all the clubs, perhaps even put together. All of them have active public websites presenting a variety of topics including breed standards, breed history, virtues of character and taste, what to look for when buying, how to pedigree identify, pedigree stock and fattening weaners for sale, pedigree meat suppliers, lists of members, husbandry advice, showing guides, membership details and much more. So if you wish to know more about a breed or are thinking about making a purchase, then they're all worth a visit.

If you're considering breeding the British Lop or Kunekune, you will need to join the relevant breed club to be a registered breeder. These two breeds govern their own herd book registrations, and to exhibit your pigs under the 'club rules' your pigs will need to be registered. The British Lop Pig Society (BLPS) also issues pedigree meat certificates for those pigs that don't reach the required breed standard, and the BKKPS uniquely provide 'pet' registrations for non-breeding stock which are eligible to be shown and are considering promoting the use of pedigree meat certificates.

Advantages to the Specific Pig Breeds

The breed clubs have a close affiliation with the BPA, usually via the BPA Breed Representatives, and importantly use the activities of their club to underpin the conservation and preservation aims of both the BPA and the Rare Breeds Survival Trust (RBST) where applicable. A recent example of this in late 2013 was identifying the last remaining British Saddleback boar of the Dominator bloodline being used by a lapsed BPA member in a cross-breeding programme. Not wishing to lose this bloodline forever to history, the British Saddleback Breeders Club (BSBC) arranged for the delivery of unrelated British Saddleback sows and pre-purchased six boar progeny to secure the bloodline, to be sold to BSBC members across the UK.

Each breed club markets and promotes its breed online and at agricultural or farming events using a website as the public face, backed up with promotional banners, flyers, stickers for pedigree meat and information leaflets for distribution. These are often co-financed between the breed clubs and the RBST, such is their importance.

So What Advantages Does it Give You as an Individual?

Most breed clubs issue regular newsletters and/or yearbooks keeping you up to date with the latest news, information on training courses available, breed standard workshops (Figure 7.1) artificial insemination, grants available, club contacts, showing news and any new rules and regulations from the government or BPA/BLPS/BKKPS that are relevant. These are often sent out electronically and as hard copies for the computer phobic so no one misses out.

Regional groups are often supported by the main club and so there is a highly social element where 'tip sharing' and support is commonplace. National meetings are held once a year at the AGM so you can meet up with old friends and be introduced to new ones and vote on how you wish *your* club to be run. The social aspect is also rapidly entering the 21st century with active Facebook and Twitter accounts, which are supervised by club committee members to maintain quality and all have the shared aim of promoting the breed and helping fellow breeders by sharing common experiences. These are open to all types of members, including enthusiasts without pigs, who usually only pay a nominal fee to join.

Breed club websites are often the first port of call for any prospective purchaser and as a member you can be listed as a registered breeder in your area and advertise your

FIGURE 7.1 A Tamworth Breeders' Club workshop in progress.

Photo courtesy of Liz Shankland.

pedigree meat for sale, boars for hire or breeding, fattening stock for sale, plus often any piggy-related services you offer, e.g. pregnancy scanning, private training courses or livestock transport. Generally, club websites have members' forums as well, so any adverts can be placed online and updated personally as required, but seen by anyone, so be sure to include full contact details.

A club may have a strong following of prospective purchasers from overseas. Many have paid good money to import our pigs and become members to keep up with the latest news from the UK. So joining opens up an additional market of potential customers, which in these times is not to be sniffed at.

You can use club logos on your own websites to show your prospective customers that you have an affiliation to the club, increasing the confidence in your livestock. Some clubs require you to ask for written permission first, while others allow open use if your subscription is current.

For the showing fraternity, all breed clubs provide Breed Champion rosettes for members who exhibit at national agricultural shows, including all the BPA accredited and affiliated shows. Additionally the majority host a Champion of Champions final, when all breed champions that year compete for the coveted title. Some clubs also host other awards, e.g. the Gloucestershire Old Spots Breeders' Club have a yearly award for GOS Pig of the Year, and also Northern and Southern Regional Championships. Another

TABLE 7.1 UK breed clubs contact details.

BREED	TYPE	RBST WATCH LIST	OFFICIAL WEBSITE	SOCIAL MEDIA FB /TW	CLUB FEES (2016)
Berkshire	Traditional and rare breed	Vulnerable	www.berkshirepigs.org.uk	FB TW	£20–£25
British Saddleback		At Risk	www.saddlebacks.org.uk	FB TW	£5–£25
Gloucestershire Old Spots		Minority	www.gospbc.co.uk	FB	£10–£25
Large Black		Vulnerable	www.largeblackpigs.co.uk	FB	£15–£20
Middle White		Vulnerable	www.middlewhite.co.uk	FB	£20
Oxford, Sandy and Black		At Risk	www.oxfordsandypigs.co.uk	FB	£15
Tamworth		Vulnerable	www.tamworthbreedersclub.co.uk	FB TW	£15
British Lop		Vulnerable	www.britishloppig.org.uk	FB TW	£5–£30
Welsh	Modern and rare breed	At Risk	www.pedigreewelsh.com	FB TW	£10–£25
Kunekune	Imported	N/A	www.britishkunekunesociety.org.uk	FB	£14

valuable benefit provided by some of the clubs is £2m public liability insurance for pigs being exhibited, even if you're just attending a local school or fete. Insurance is becoming increasingly demanded by event organisers, and the club insurance provided is accepted as valid.

All this for a modest yearly fee . . . it's a no brainer, as they say!

REGISTRATION AUTHORITIES IN THE USA

The USA has quite a few societies that manage the registration of the specific breeds.

The National Swine Registry governs the identification and registration of pedigree Yorkshire, Hampshire, Landrace and Duroc. The Certified Pedigreed Swine Association governs the Chester White, Poland China and Spotted Swine. The American Berkshire Breed Association governs the Berkshire breed.

The identification and registration of heritage breeds are governed by the Large Black Hog Association or North American Large Black Pig Registry, the Tamworth Swine Association, American Mulefoot Breeders Association or American Mulefoot Hog Association and Registry, American Guinea Hog Association, Red Wattle Hog Association and the National Hereford Hog Association. The Livestock Conservancy is currently responsible for the official recording of the Choctaw and Ossabaw Island pig breeds.

Other US breeds are provided with a pedigree registration service from the American Mangalitsa Breeders Association, the American Kunekune Pig Society (largest Kunekune Pig registry), the American Kunekune Pig Registry and the Juliana Pig Association and Registry.

The Gloucestershire Old Spots, the Mulefoot, Kunekune and the Large Blacks have a choice of two registration bodies. I suspect politics has played a role somewhere down the line, and so read up on/ask other breeders about both groups before deciding which one to go with. When I asked some American breeders of those breeds, it was obvious which registries were inefficient, so ask around – I obviously cannot write what they all said. Also one of the groups might not recognise the pedigree of a pig registered with the other group, so check with your breeder you purchased the pigs from, as you may not get a choice.

In all cases, to become a registered breeder, you must join the appropriate breed registry. You will be asked to come up with a unique herd identifier (herd mark), which is a combination of two to four alpha-numeric characters. This becomes the prefix of all the pigs you register in the future.

SOURCING PEDIGREE BREEDERS

The easiest way is to either look at the specific breed society's websites, in specialist smallholder magazines or on the BPA/TLC/Breed Association websites. Some breeders will be listed on a non-registry specific breed's club website as well, but all breeders of registered stock have to be full members of the appropriate breed registration body, so you can look them up and decide how far you are willing to travel.

All pedigree litters born will be pre-registered (birth notified UK and USA or Litter Certificate USA) with the appropriate registering organisation. Not all will go forward for full registration, only those that strictly adhere to the breed standard. The piglets that are eligible for full registration will already be identifiable by their ear markings if they have been weaned and from seven days old in the USA. Depending upon the breed of pigs, some will have notches cut out of their ears which denote their number, others will have the number tattooed in their ear(s) and some will have ear tags. In the UK the pigs will also carry one ear tag with the breeder's HDL code printed on it in addition to the notch or tattoo. UK breeders who don't wish to notch or tattoo, or cannot because of Freedom Food membership, may obtain permission to double ear tag their pigs to maintain their pedigree status, but these pigs are not eligible to be exhibited at BPA-accredited shows. Ear tags can be plastic or metal as long as they are heat resistant.

The pedigree pigs you choose may *not* have full registration yet, especially if they are still young, but this will be done by the breeder, and the herd book registration certificate will be sent directly to you and be put online. In the USA, the breeder should give you a signed Litter Certificate and upon joining the registry you can complete to full registration. Older pigs may already be registered, in which case the breeder will transfer ownership to you and again you will receive a copy of the updated herd book registration certificate. I usually do this in a spectacular fashion at the time of sale online to give an added sense that they are buying something special . . . which they are!

BREED STANDARDS/STANDARDS OF EXCELLENCE

All pigs registered for breeding must achieve the minimum breed standard for their breed in order to be suitable, and should preferably exceed it; the remainder of the litter should be put in the fattening pen. When going to look at pedigree pigs, carry a copy of the breed standards with you until you have a trained eye. These can be downloaded free of charge from the UK and USA registration authorities' websites or specific breed club websites, or ask your breeder or course provider to supply you with one. You will get to know the look or bloodline that you like and will be able to pick the individuals that you wish to select for closer inspection. However, it is worth re-inspecting at regular intervals, especially the teats, as these can sometimes invert, and although you can show any registered pig, it will not win with such an obvious fault. Generally across the breeds, to achieve herd book registration the pig (boar or sow) needs to have: at least 12 evenly spaced nipples, preferably 14 or even 16; been bred in the UK/USA; birth notified/Litter Certificate obtained from registered parents and have the correct ear markings applied.

If you are in any doubt, most breeds have a breed representative/ advisor or an area regional representative who will be able to help you. The ones in the UK are listed on the BPA, British Lop and Kunekune websites. Also many breeds have breed clubs, which also have advisors, and again these can be contacted and in many cases will inspect the pigs for a modest mileage charge.

How Do I Know My Pig is a Registered Pig?

Every pedigree pig eligible for registration will have an identifiable ear marking. If it doesn't, it is just a pig and not a pedigree pig.

EAR MARKINGS IN PEDIGREE PIGS (UK)

Tattoo ear marking is used for Gloucestershire Old Spots, Large White, Middle White, Tamworth, Welsh, British Lops and the Mangalitza. It may also be used for Durocs and Oxford, Sandy and Blacks in preference to notching but must be declared at the time of the first birth notification.

The females are marked on both ears, with their number placed inside on the right and outside on the left, as you look from the rear. Stud males can additionally have the breeder's HDL in their right ear and the numbers on the left, or both on the left ear. In addition an ear tag for movement purposes *must* be applied. This can be either metal or plastic and must show the herd mark on one side and the HDL and individual pig number on the other. The herd mark is nothing to do with pedigree status and is the same as you would use for meat weaners going to slaughter, except that it is the breeder's herd number, not yours. Pedigree pigs have a special derogation to be able to be moved without having to retag at each holding.

Ear notching is used for British Saddlebacks, Berkshire, Duroc, Hampshire, Oxford, Sandy and Black and Large Black breeds (Figure 7.2). All the breeds have the same notch numbering system except the British Saddleback, which has its own. The notches are the pig's primary identification and the ear tag is for supplementary and legal information only, i.e. the HDL code and your herd number. What is also allowed is that if the British Saddleback breeder does not want to perform the 400 notch in the middle of the left ear, then when the number 399 is reached the number can go back to number 1 but with 'HDL A' on the ear tag. When 399A is reached, it can go back to notch 1 with 'HDL B' on the ear tag, and so on.

Ear tagging is used to identify British Kunekunes and they have their very own BKKPS ear tags with a special pig emblem on them as well as the legal herd mark requirements, although ordinary ear tags are acceptable as pedigree identification.

Rightly or wrongly, mostly wrongly, we only notch or tattoo the potential breeding stock, which makes it is easier to identify them when they are in a group with those

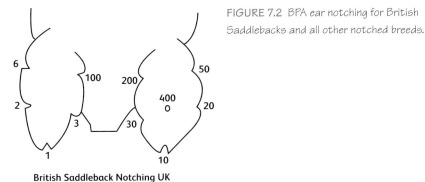

FIGURE 7.2 BPA ear notching for British Saddlebacks and all other notched breeds.

British Saddleback Notching UK
Each location can only be notched once.

All other UK notched breeds.
Use the least notches to make up the number.
Each location may be notched twice if required.

Units   1000's   100's   10's

destined for fattening. To maintain pedigree status in order to obtain the pedigree meat certificate, you are supposed to notch/tattoo and ear tag every pig in the litter, but I cannot see the point in creating a wound or distressing piglets when they will be eaten in a few months. They still get allocated a number in my pedigree records, but I don't do anything to the actual pig. If you wish to pedigree identify correctly, then all pedigree piglets, whether for registration or fattening, should carry the full pedigree identification.

If the pigs lose their tags or they need replacing due to breakage or fading, permission is required by the registration body. If you have chosen the double-tagging system, you must do this quickly in case the other tag is lost. If the pig has no means of identification, then extensive blood testing may be required, at your expense. The ear numbering system makes each pig uniquely identifiable until it dies and beyond.

EAR MARKINGS IN PEDIGREE PIGS (USA)

The many different breed registries in the USA have different demands for pedigree identification and some allow the breeder to choose. Tables 7.2 and 7.3 show the permissible options for each breed. USDA-approved ear tagging and tattooing is previously described in Chapter Two.

TABLE 7.2 USA pedigree ear-marking requirements of modern breeds.

MODERN BREED	REGISTRY	NOTCH SYSTEM	USDA '840' EAR TAG
Duroc	National Swine Registry (NSR)	Universal Ear Notching (UEN) Max. 161 litters	Additionally required when exhibiting at ABA, CPS or Team Purebred affiliated shows
Hampshire			
Landrace			
Yorkshire			
Chester White	Certified Pedigreed Swine (CPS)		
Poland China			
Spotted Swine			
Berkshire	American Berkshire Association (ABA)	Modified UEN system Max. 1199 litters	

TABLE 7.3 USA pedigree ear-marking requirements of heritage and other breeds.

HERITAGE AND OTHER BREED	REGISTRY	NOTCH SYSTEM	USDA '840' EAR TAG	USDA-APPROVED TATTOO	ADDITIONAL OPTIONS
Choctaw	The Livestock Conservancy (TLC)	TLC is initiating a formal conservation program. See TLC website for current situation			Largely a free-roaming breed
Ossabaw Island Pig		Currently (2016) undergoing an Ossabaw Hog Recovery Program with recovered pigs being ear tagged and photographed			
Gloucestershire Old Spots	Gloucestershire Old Spots of America (GOSA) Gloucestershire Old Spots Pig Breeders United (GOSPBU)	Universal Ear Notching (UEN) system Max. 161 litters	Or ear Tag Can be colour coded for management purposes	Or tattoo	Two photos (left and right)
Guinea Hog	American Guinea Hog Association (AGHA)	UEN system or own system	Or ear tag		Other permanent ID permitted
Hereford	National Hereford Hog Association (NHHA)	UEN system			

HERITAGE AND OTHER BREED	REGISTRY	NOTCH SYSTEM	USDA '840' EAR TAG	USDA-APPROVED TATTOO	ADDITIONAL OPTIONS
Juliana Pig	Juliana Pig Association and Registry			Tattoo in right ear	Or microchip Three photos
Kunekune	American Kunekune Pig Society (AKKPS) American Kunekune Pig Registry (AKKPR)		Ear tag	Or tattoo	or microchip
Large Black	Large Black Hog Association (LBHA) North American Large Black Pig Registry	UEN system	Optional in addition to notch		
Mangalitza	American Mangalitsa Breeders Association	UEN system			
Mulefoot	American Mulefoot Breeders Association American Mulefoot Hog Association and Registry	UEN system	Or ear tag		
Red Wattle	Red Wattle Hog Association	UEN system	Or ear tag	Or tattoo	
Tamworth	Tamworth Swine Association	Modified UEN system Max. 80 litters			Renewed litter numbering every year

Universal Ear Notching (UEN)

Pigs that are pedigree notched in the USA follow the same universal ear-notching (UEN) system, with the exception of the Berkshire and Tamworth breeds, which use modified UEN systems. The American Guinea Hog Association also gives you 'This MUST be performed before seven days of age to maintain pedigree status.' The pig's right ear as viewed from the rear of the pig is designated as the 'litter' ear. Notches in this ear tell the breeder which litter the pig was born to.

NOTE: Breeders may be required to start with litter number 1 on the first of the year or the first of a specific month, start at 1 and keep going until all 161 litter numbers have been born, start at any litter number they like and in any order or restart the litter numbers twice per year – depending upon the regulatory requirements of the individual registration body.

FIGURE 7.3 *The universal notching system (UEN) used in the USA.*

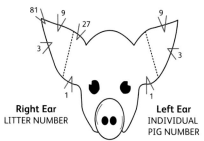

UNIVERSAL EAR NOTCHING SYSTEM (UEN)

Right Ear
LITTER NUMBER

Left Ear
INDIVIDUAL
PIG NUMBER

EACH NOTCH MAY BE USED TWICE EXCEPT
THE 81 NOTCH. A FEW EXAMPLES BELOW;

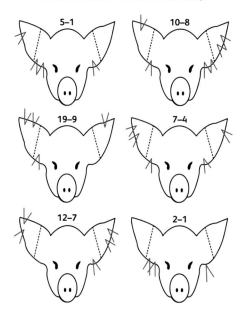

All litter mates born to the same sow have the name litter number and will have identical notches in the right ear. The pig's left ear designates the pig number within the litter and so starts at 1 with each litter born. So, the first piglet would be pig number 1, the second piglet is pig number 2 and so on. These individual numbers are randomly assigned to piglets and not associated with sex or birth order. A notch of 14 in the right ear and 6 in the left ear means it is the 6th pig out of the 14th litter born on the farm and would be written as 14-6.

Notches are only allowed a maximum of twice at each number location, except the 81 notch, which is only allowed once (Figure 7.4). Use the minimum amount of notches to achieve the desired number. The right litter ear has a maximum of 161 litters able to be recorded (81 + 27 + 27 + 9 + 9 + 3 + 3 + 1 + 1 = 161) and the left piglet number ear has a maximum of 26 piglets (9 + 9 + 3 + 3 + 1 + 1 = 26), which is more than adequate as it's highly unlikely you will have a litter of more than 26.

Modified UEN Systems

The Berkshire system (Figure 7.4) has been modified to use the left ear for additional litter numbers and up to 1199 litters can be accommodated. The 9, 3, 1 system in the left ear remains the same.

The Tamworth notching system (Figure 7.5) is the same as the UEN system but with the 81 notch not used. Here only 80 litters can be accommodated.

All notching systems are used in conjunction with the individual pig's Litter Certificate, birth notifications or full registration documents, which denote the parentage and date of birth – you will need the notched pig and the Litter Certificate/birth notification certificate to enable registration or registration certificate transfer of a previously registered pig. Once registered, each pig will have its own uniquely numbered registration number linked to its ear notch number and year of birth.

Berkshire Ear Notching System
Litter Mark

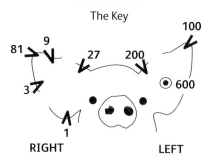

Litter Mark: Right and left ears are used for litter mark. All pigs of the same litter must have the same ear notches. Right ear is the pig's own right.

Individual Pig Marking

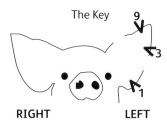

Individual Pig Marks: Left ear is used for notches to show individual pig numbers in the litter. Each pig will have different notches in this ear.

FIGURE 7.4 *The Berkshire notching system (from the breed registry website).*

TAMWORTH EAR NOTCHING SYSTEM

Guide for showing location and value of ear notches for marking litters and individual pigs, Breeders may change over to this system in 1966.

THIS IS THE ONLY SYSTEM WHICH WILL BE PERMITTED AFTER JANUARY 1, 1967
Number litters consecutively for one calendar year, Start over with Number 1 at the beginning of each year. The Association may refuse to register or certify for showing any animal not properly ear marked. All pedigrees sent in for recording must be made out on the blanks of this Association.

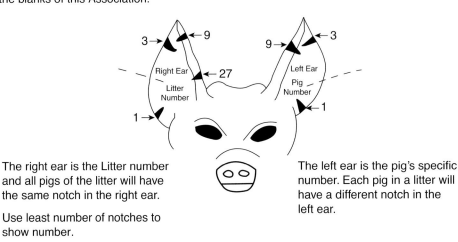

The right ear is the Litter number and all pigs of the litter will have the same notch in the right ear.

Use least number of notches to show number.

The left ear is the pig's specific number. Each pig in a litter will have a different notch in the left ear.

FIGURE 7.5 *The Tamworth notching system (from the breed registry website).*

A PIGLET'S NAME

In the UK, a piglet's name is made up of the prefix of the breeder, followed by the sow or boar bloodline name, and then the number of the piglet. The bloodline name for boars is taken from the sire and for gilts it's taken from the sow. Your very first litter of a breed, not bloodline or individual sow, will start at 1, but subsequent litters from the same breed carry on from the last number used in the previous litter. The boars in each litter are numbered first, followed by the gilts. Our breeding prefix is Tedfold, so our first boar piglet from our British Saddlebacks born on our farm was from a purchased sow named Timberline Molly 506 mated with boar named Framfield Consort 741, and was named Tedfold Consort 1. As there were five male piglets in total, these were numbered 1 to 5 and then the first gilt was called Tedfold Molly 6 (Figure 7.6).

If you then choose another breed of pig, then you can start at number 1 again. This can be confusing on the paperwork, so pay attention to keeping it correct. For example, both the British Saddleback and the Middle White breeds have the male bloodline Rajah. So we could feasibly have two pigs on the farm called Tedfold Rajah 1, two called Tedfold Rajah 2, etc.

In addition to the pig's name, it also has a unique registration number (R Number) allocated at the time of registration, which includes a two-letter breed suffix. So in our case of the two Rajahs, one would have the suffix BS and the other MW, e.g. R123456BS or R123457MW. Of course looking at the pigs will tell you who is who as well.

In the USA, the naming of a piglet is not uniform across the breeds and it is imperative that you consult your individual breed registry to understand what you are required to do. Listing all the different requirements for each registry isn't really feasible. All pedigree pigs put forward for breed registration will be allocated a unique breed registration number.

Some registries will name in a similar way to the UK, with the breeder's prefix of either the farm name or breeder's two-four alpha-numeric prefix, the bloodline name with gilts named after the mother and boars named after the father, and then the litter number, e.g. 14-3. Others rely upon the permanent ID requirement, e.g. tattoo of the registration number, and the name can be anything you choose. In some ways, the latter option is preferable as in the UK, where bloodline names are always used, breeders can be led to believe that if a pig has a different bloodline name to another, then they are unrelated, whereas in reality they could share a considerable amount of genetic heritage, even more than one with the same bloodline name. What breeders should

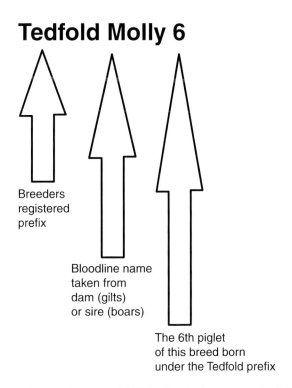

Tedfold Molly 6

Breeders registered prefix

Bloodline name taken from dam (gilts) or sire (boars)

The 6th piglet of this breed born under the Tedfold prefix

FIGURE 7.6 The name of the sixth British Saddleback piglet born, using the Tedfold prefix.

TIP

Always know exactly how to identify your piglets before you start notching and tattooing. Both are hard to correct if you get it wrong.

be using is the calculated Co-efficient of Inbreeding data, collated at the time of registration, when selecting potential mates, and almost ignoring bloodline names (see Chapter Eight).

You do not need to own the boar you use in order to be able to name it as the sire. In the UK the boar owner will get a notification you have used their boar, but in the USA you will need an AI certificate or a Certificate of Service/Breeding.

NOTE: Your specific breed may have a choice of different naming and numbering systems, so contact your breed registry for further advice, e.g. Gloucestershire Old Spots has two different ways of recording their breeding (normal and cyclic) in the USA – it is now discouraged in the UK with the Co-efficient of Inbreeding analysis preferred. In the UK there is also an alternative BPA numbering system for all breeds where the pig's registered name doesn't match the number of piglets born, and goes up in accordance with the number registered within the herd. In some ways this is better for keeping track of those considered good enough to enter breeding programmes and the allocation of a unique registration number remains.

DNA TESTING BEFORE FULL REGISTRATION (USA)

In the USA, some breed registries require a DNA sample, which may be kept for record or may be used for testing for parentage and/or to determine the presence of one or more of the stress genes, e.g. Halothane gene, Rendement Napole gene, both of which can potentially affect the quality of the meat (see Chapter Eight). Where it is required, the sample must be submitted before a registration number is allocated. For parentage testing it is usually a hair sample with hair roots, and for stress gene testing it is blood or semen put onto a special blotter card.

Most USA-registered AI boars will have to be stress gene tested as negative before their semen is allowed to be sold, and this is listed on their pedigree registration certificate, e.g. HAL1843 *neg* or *nm*.

Additionally at some shows and exhibitions all pigs including the barrows are required to be stress gene tested in order to be exhibited, and to ensure meat quality some premium meat schemes also require stress gene testing.

UNDERSTANDING THE BASIC GENETICS OF BREEDING AND INHERITANCE

One aspect of genetics is the science of heredity, and like all 'genetics' it is controlled by genes. Genes are DNA-coded sets of instructions to inform the body on every functioning aspect, from visual (phenotypic) appearance to minute cellular instructions. There are around 21,000 genes within an individual pig and, by and large, all works well.

In straightforward Mendelian inheritance (Figure 8.1), each normal gene carries two copies, one donated from each parent, and they operate together with one as 'back up'. Traits, both good and dysfunctional, may in some instances be carried on multiple genes but usually they are carried on a single gene. If on a single gene both copies do not match, then whether you see it phenotypically or can detect it in the pig depends upon whether it is a dominant gene, a recessive gene or an X-linked recessive gene. A pig with a dominant gene will display the particular trait; one with a recessive gene may display the trait or be a carrier; and one with an X-linked gene will be a carrier or affected, depending upon which parent was the carrier or whether it was both parents. Although both desired and undesired characteristics are inherited the same way, it has the most significance when it's a defective version of the gene that is passed on, and using a defective gene as an example makes understanding inheritance a little easier. When one of gene pairs is dysfunctional or undesired, e.g. prick ears in a lop-eared breed or an extra cley on the trotter, its ability to be passed on to subsequent progeny depends upon what type of gene it is and the status of the corresponding gene in the other parent.

GENE INHERITANCE

Dominant Gene Inheritance

If one parent has one of the pairs as a dominant dysfunctional gene and the other has a double normal gene, then there is a 50 % chance of each piglet born displaying the trait. If both parents have one of the pairs as a dominant gene, then there is a 75 % chance of each piglet displaying the trait. Of course most breeders would not use a pig with such an obvious defect for breeding but it can still happen with unseen conditions, e.g. heart murmur, diabetes, enzyme deficiencies, etc.

Recessive Gene Inheritance

When one parent has one of the pairs as a recessive gene, then the pig may be unaffected or be a carrier. Carriers are perfectly healthy as the correct version of the gene is being used, but they carry a copy of the defective gene as the 'back up' copy that may arise in subsequent progeny if they are mated with another pig that is also a carrier of the same defective gene. If a carrier pig is mated to another carrier pig, then each piglet has a 25 % chance of not having the gene, a 50 % chance of being a carrier of the gene and a 25 % chance of exhibiting the trait. If the carrier was mated with a pig without the defective gene, then each piglet has a 50 % chance of being normal and a 50 % chance of being a carrier – but 100 % would look outwardly healthy and phenotypically free of the particular trait.

In the national herd, recessive gene carrier status may build up unseen over time and realistically achieve 20 % of pigs in a total population before the trait started to show itself by producing double recessive gene pigs which do show the undesirable trait. Of course, 20 % may be achieved within the same herd quite quickly if genetic diversity isn't maintained, especially if inbreeding or line breeding is practised.

Sex-Linked Gene Inheritance

To confuse the matter a little more, some genes are donated in groups with other genes. If a defective gene is transferred with the same gene that determines each piglet's sex, it is known as sex-linked or X-linked, as these transferred genes are typically on the X chromosome. Gilts always have two copies of the X sex gene (XX) and boars always have one X from the sow and a Y from the sire (XY).

- For X-linked dominant genes in the sow, the gilts and boars have a 50 % chance of being affected. If the sire carries the gene, then all the gilts and none of the boars would be affected. This is because the gilts will always inherit their father's X chromosome.
- With X-linked recessive genes carried by the sow, then the gilts have a 50:50 chance of being a carrier or unaffected and each boar piglet has a 50:50 chance of being

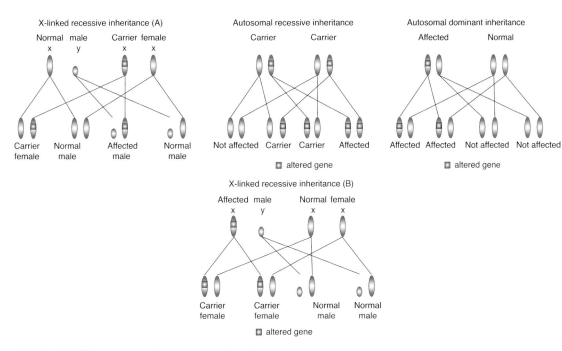

FIGURE 8.1 *Different types of parental gene inheritance.*

affected by the trait or unaffected. If the gene is carried by the sire, then all the gilts will be carriers and the boars will be unaffected.

The hardest to determine in a population are the recessive inherited traits, and extensive population DNA testing of case, case control and non-case pigs is required to formally detect it.

WHY DOES INBREEDING MATTER?

As you have seen, passing on undesirable traits is directly linked to a pig's breeding. The more closely the parents are related, the more frequently these undesirable traits will result in the trait becoming phenotypic. Line breeding uses exactly the same principles but a desired characteristic is being sought; this is a game best played by others. For most breeders, genetic diversity is the key to producing healthy litters and robust examples of future breeding stock. This is not always easy when the genetic pool to pick from is small, but it is very important. Luckily, with modern technology, not only can you find out on a computer how related one pedigree pig is to another, you can also use artificial insemination from boars all around the world to assist you in lowering the risk – once you know what you are doing.

The biggest mistake some breeders make is registering lots of boars from the same litter to use as sires. One should be the maximum number from each litter or you risk producing lots of boars that are all highly related to be used in breeding programmes, pushing up the number of litters that are closely related and not expanding genetic diversity at all. Use different unrelated AI boars each time, remembering that other breeders will be using them as well – making some of your herd highly related to others around the country. AI stations do change their boars or add new ones to the list, so take advantage. It is recommended for breed conservation that one individual male should not contribute to more than 5 % of future breeding populations, and so the most popular AI boar may not be the best way forward.

CO-EFFICIENT OF INBREEDING (CI)

In the USA you will often see a pig's Co-efficient of Inbreeding (CI) quoted as a percentage and in the UK it is quoted as a score between 0 and 0.1. The CI is the probability that a pig with two identical genes received both genes from one ancestor and not implicitly that an individual did actually inherit that identical gene. The more related the two pigs mated are, the more likely they will have identical inherited genes. The CI gives a guide as to how related a pig's parents were to each other and can be used prior to a mating to calculate the percentage of inbreeding any subsequent piglets would have. The lower the percentage/score of inbreeding, the more likely the piglets will be healthy. As piglets in a litter are not clonal (unless identical twins) and all genes are donated randomly from each parent, CI percentages/scores can only be used as a guide and not an absolute. They are a useful additional tool to use when selecting between equally suitable boars to use, and in assessing the purchase/selection of future breeding boars or sows and/or their potential increased susceptibility to suffer from unseen health problems such as sub-fertility, or congenital abnormalities in the piglets.

In its most simplistic form, use the example of a herd with no history of inbreeding, then the mating of first cousins who share one set of grandparents. For any particular gene in the male, the probability that his female cousin inherited the same gene from the same grandparent is 12.5 %. The probability that the sow will pass the same gene to her progeny is 50:50 (0.5) so 12.5 x 0.5 = 6.25. Thus the piglets from an otherwise outbred herd from a cousin mating has a CI of 6.25 % or a score of 0.06.

If you now start adding in greater degrees of ancestral inbreeding, especially if the same common ancestors are shared more than once and on both paternal and maternal sides, the probability of inheriting identical genes increases and therefore the CI percentage/score starts to rise considerably. The impact of recent inbreeding in the previous two to four generations is also greater than early inbreeding followed by outbreeding. It is this last point that can be used to recreate genetic diversity back into your herds by using partners that will produce the lowest CI percentage/score in subsequent litters.

Breeders should always aim for the lowest CI possible in the subsequent piglets when selecting the parents and accept that it may take a few generations to get down

TABLE 8.1 The minimum Co-efficient of Inbreeding percentage of a related mating with no previous inbreeding history.

SOW–BOAR RELATIONSHIP	AVERAGE % OF SHARED DNA	CI % (USA) OF PROGENY	KINSHIP CI SCORE (UK) OF PROGENY
Identical twins (impossible but useful to demonstrate the 50:50 of genes donated randomly)	100%	50%	0.5
Full siblings Parent–child	50%	25%	0.25
Half siblings Aunt/uncle Niece/nephew Grandparent–grandchild	25%	12.5%	0.13
Great grandparent–great grandchild First cousins	12.5%	6.25%	0.06
First cousins (once removed)	6.2%	3.1%	0.03
Second cousins	3.2%	1.6%	0.02
Second cousins (once removed)	1.6%	0.8%	0.008

to lower levels. In an ideal world a CI of 0 % should be aimed for, but in a limited pool of pure-bred pigs this may prove impossible, and if you can get it lower than 10 % in the USA, then you're doing well. In the UK, where individual pedigree breed populations tend to be in greater numbers and the inbreeding level is only an average of 5 to 7 % nationally, it is recommended that planned matings should be a maximum of 0.02 (CI 1.6 %) to continue to improve our breed diversity.

Founders

Some conservation geneticists maintain that it is desirable to have representation of 'founder' animals in the pedigrees of the latest generation in order to maintain genetic diversity. However, it is impossible to go back to the true 'start' of a breed and 'founders' are usually defined as those animals that had progeny but no parents, in a database which gives an arbitrary start date for a breed, and so the CI calculations will be flawed if the start dates for particular bloodlines vary, e.g. the CI % with a start date of 1980 will use different 'founders' than a start date of 1990. In the USA it is important to analyse the same number of generations when calculating and comparing CI %. The UK uses a different starting point for their kinship CI calculations and the figures provided already take this into account.

Outliers

If you have found yourself with a high CI% sow without experiencing any symptoms of inbreeding (aka Depression of Inbreeding), then consider yourself lucky, but the higher the pig's CI%, the more you need to find unrelated partners if you wish to carry on breeding pure. As CI% is calculated from a series of probabilities rather than absolute measurements, it may be that she had luck on her side with regards to inherited identical genes and is what's known as a 'probability outlier' so culling isn't required. The same is true of boars, but the impact is greater as boars can be used on multiple sows, so I would still pop that chap in the fattening pen.

ACCESSING CI% INFORMATION

Where the service is available, contact the registrar of your breed registry/online members' facility (USA) or your Kinship analysis representative for your breed in the UK and they will be able to calculate the CI% or the CI score that would be in your potential litters.

This is a massive topic and what is written here is just to let you know this facility exists, how it should be used and a bit of theory to help you understand its importance.

DNA TESTING FOR THE STRESS GENE

When testing for the stress gene, it shows the actual presence or absence of the gene rather than a probability. The same basic Mendelian inheritance rules apply but the gene has been identified and can be specifically tested for in the lab.

If you know the gene status of the parents already, you can predict the outcome in the progeny (Figures 8.2–6). For ease of understanding, I have called stress gene negative pigs ss, the stress gene carriers Ss and stress gene sufferers SS, remembering that only one gene from each parent is donated.

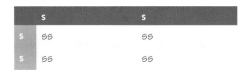
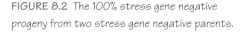

FIGURE 8.2 The 100% stress gene negative progeny from two stress gene negative parents.

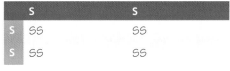

FIGURE 8.3 The 100% stress gene positive progeny from two stress gene positive parents.

	S	s
s	Ss	ss
s	Ss	ss

	S	S
s	Ss	Ss
s	Ss	Ss

FIGURE 8.4 *The 50% negative/50% carrier progeny from one stress gene negative parent and one carrier parent.*

FIGURE 8.5 *The 100% carrier progeny from one stress gene positive parent and one stress gene negative parent.*

	S	s
S	SS	Ss
s	Ss	ss

FIGURE 8.6 *The 25% positive, 50% carrier and 25% negative progeny from two carrier parents.*

In some USA breed registries and for some activities, e.g. showing or premium meat schemes, stress gene testing is mandatory. In the UK, for breeders of pedigree and/or traditional breeding, testing isn't currently required.

CAN YOU BREED PIGS AND WORK FULL-TIME?

Quite simply, yes you can. Be fair to the pigs though and don't start breeding until you have some experience of keeping pigs. If you don't have an experienced mentor on hand, then give it even longer. You will make a few mistakes along the way even if you have been on a pig course or own every book written about pigs, and if you get it wrong with a pregnant sow or a breeding boar the results could have far-reaching ramifications – to you, to your pocket, but most likely to the pigs themselves.

It is worth doing your homework and choosing your breed carefully, as some pigs are known for their excellent mothering ability and docile, easy-to-handle temperaments – see Chapter Four for an overview of the different breeds. If you can book leave or arrange for help to be available during farrowing time, then so much the better.

I would choose a breed that is contented with its lot in life and less likely to challenge your fencing. Choosing one with a good mothering ability means you will have more live piglets at weaning time to sell or rear for meat; you will spend less time trying to hand rear less robust piglets (seriously, you will not be able to do this if you work full-time, let alone commute); the piglets you do get will be healthier from the high-quality milk and the sow will be able to do most/all of it on her own without human interference and vet bills. I think it's called a 'no brainer', when time is your most important factor.

> **TIP**
>
> When time restricted, choosing a breed that is easy to handle makes your day-to-day activities simpler and therefore less time consuming.

BREEDING FROM YOUR PIGS

If you want to breed, you are going to have to consider how you will get your gilt or sow pregnant. If you wish to start breeding without addressing this issue for a while, then consider buying an already pregnant or 'in-pig' gilt or sow. You will pay more than you would for a weaner but your costs will probably be wholly or partially negated by the piglets she is carrying, either by selling them live at eight to ten weeks or rearing them for meat yourself. Expect to pay more the 'closer she is to profit', which, in other words, is her due date.

Remember that even when pregnant she will need the company of another pig, so either purchase two pregnant gilts/sows or a weaner or two for company in an adjoining pen. You can always eat the companions when she is due to farrow.

I would buy at least two companions, as I will only send pigs to slaughter in pairs or more, so they have some comfort during the experience by being next to a chum at all times. The advantage of buying a pregnant gilt is that she is young and you will get a fair few years of breeding litters out of her; traditional pigs kept by smallholders tend to stay the duration of their lives on the same farm and she could breed up to six years old or even longer if you are lucky and you keep her in good condition and health.

The disadvantage is that you have no idea how she is going to react to her first farrowing. She may be terrified by the whole experience and even attack her babies. The litter size is generally smaller than with an established sow and it is said that you should never select future breeding stock from a first or even second litter. I personally don't agree with avoiding the second litter necessarily, as why would pigs be that different from every other mammal out there? That could be applied to the first litter as well but I can see some logic here, in that there was less room for the piglets to grow, as she was younger.

If the piglets of second and subsequent litters are of breed standard and grow well, then fine. If you buy a sow that has had one litter already and is pregnant again, then you know she is likely to conceive again and she will have a fair idea of what she is doing at farrowing time.

The number or quality of piglets in a litter generally increases as she gets older. There are two schools of thought I have heard about litter size and age; the first is that litter size peaks when the sow is around three years old and then slowly decreases again; the second is that she continuously releases more eggs as she gets older and so the litter size constantly increases however, the piglet size gets smaller and the number of runts, non-viable piglets and reabsorption rates increase, resulting in a lower number of live piglets. Either way it would seem sensible to me to buy an 18-month-old sow that has had one litter already but is still on the way up in terms of number of viable piglets per litter. As expected, these are more expensive than maiden gilts for the reasons stated.

We wished to have certain bloodlines of our British Saddleback pigs and so all bar a couple were bought at eight to ten weeks old and we got to know them intimately over time. This meant that when we finally did get them pregnant and they farrowed, we were, without exception, allowed to be at the birth and even to cuddle the new piglets after a few days without the mums getting upset. We did have to accept, however, that it would take us much longer to establish our herd and we ran the risk of the females not being fertile. Of the two pregnant gilts we bought, one let us handle her piglets and attend the birth, but the other was less keen and only allowed my husband to go in with her; I was forcibly removed from the area. A thought to consider when buying your future breeding stock. It really depends upon what you want to get out of the experience.

At What Age Can My Gilt Become Pregnant?

Well-kept gilts of traditional breeds are safely ready for mating at around nine to ten months old, which is approximately 130 kg (285 lb) live weight, although they will start coming into season from as early as four to five months. If you let them get pregnant at such an early age, you run the risk of affecting their growth and they may never reach their full potential. I personally think nine months is still too young and you should wait until they are at least 12 months old, although other breeders would disagree instantly and it is a common age to put to a small boar. Modern breeds usually reach the 130 kg weight earlier, at between seven and nine months.

NOTE: Smaller breeds like Kunekune will of course not meet the 130 kg weight at ten months for breeding. A generally accepted age is 12–14 months, but with some smaller gilts even longer may be required.

We got caught out once to our horror when a 14-week-old boar piglet impregnated a six-month-old gilt. Who would have thought he would have been fertile at such a tender age? We now take absolutely no chances, having learned the hard way, and to our shame she is still a little smaller than she should be. I, and a growing number of traditional breeders, prefer to wait until they are 12 months before allowing access to a boar. At the other end of the scale, if you leave the gilt too long before covering, you run the risk of her going barren and becoming a giant, expensive pet or a freezerful of sausages. The latest age for going barren remains a mystery and I can only find anecdotal evidence that this occurs – we regularly wait until 14 months without apparent detriment. I am not saying it doesn't occur, but I can find no direct proof, as most research is performed on commercial pigs, and if commercial gilts don't conceive on the second time of trying, they are sausages, so no research has been conducted! I also suspect the age varies enormously between breeds and individuals.

Before You Get Them Pregnant

Think ahead to just under four months' time: you will need somewhere suitable for your sow to farrow; you will need be around for the birth, so no business trips or conferences; and you need to know when the piglets will be ready for sale. If you plan to run them all on for meat yourself, then this is less of a worry but you will need to be able to wean them between six and eight weeks after the birth and have another pen to put the sow into. If you have two gilts or sows for mating, I would, as a first timer to breeding, stagger the births so you are not inundated with piglets all at once. It will be an issue for the later sow to be on her own, so I would suggest that either she is in a pen next to the other sow and can clearly see her, or she has a couple of companion fattening weaners in the next pen. Remember the sows will need an ark each as they will not share their maternity unit easily and getting up repeatedly to chase off the other sow from the ark may cause them to squash more piglets.

FEEDING WHEN BREEDING

There is an art to feeding the working sow, and a definitive 'how to' guide on feeding the working sow would be impossible to produce, as the 'norm' varies considerably between farms, breeds, sows in a herd and even between litters from the same sow. Include other contributing factors – such as being under/overweight; kept indoors/outdoors; the weather; lactation length; creep-feeding the piglets; parity 0 or 1 gilt so still growing herself or an established sow; any supplementary feeding – and the prescription changes yet again.

The theory behind feeding the working sow is to keep her body condition score (BCS), as described in Chapter Three, constant at between 3 and 3.5 and to vary the feed quantity and/or composition to achieve this.

The practice of the sow 'feeding a litter off her back', whereby at the end of lactation the sow is often an emaciated BCS of 1.5 to 2, is now considered outdated, with research showing that it's better for the sow and the quality of her litters to maintain the same body condition year round. Performing BCS regularly means that any feeding adjustments needn't be dramatic. In addition, there are key times when feed levels may be increased or decreased for short periods to help with conception and maintaining health. This might not be what you have read in older books or heard from long-standing pig keepers, but the published research is convincing reading. Don't get too bogged down when choosing what protein level is the best, but if you have a choice, 13–16 % is a good place to start.

To help give the theory some context it's probably easier to describe a case study. A typical British Saddleback sow on our farm has two litters per year and when pregnant is fed a maintenance ration of 5 lb of 13 % protein sow rolls per day split into two feeds to maintain the condition score of 3 to 3.5. In the last three weeks of pregnancy the feed is increased slightly to 6 lb, to compensate for the final growth of the piglets in utero and assist udder development. Two days before farrowing the quantity is reduced to 3.5 lb per day to lessen the chance of a 'farrowing fever', also called Mastitis, Metritis and Agalactia (MMA). After farrowing the feed is increased to 1 lb of sow rolls per day per piglet she is feeding with a minimum of 6 lb. Weekly BCS then begins and the feed is slowly adjusted to keep her score of 3 to 3.5. Seven days post-farrowing, creep is introduced *ad libitum* to the piglets to help start taking the pressure off the sow although they only nibble at the ration for a week or two. A sow's milk production peaks at around week three, so any problems in feeding to body score come to light reasonably quickly. After weaning, if the sow goes straight back to the boar/AI, we put her on double maintenance rations (10 lb) to 'flush'; if we are skipping a heat then we put her back on her maintenance ration of 5 lb and flush the week before she goes back to the boar/AI. As soon as mating has occurred the feed levels resume at the standard maintenance ration of 5 lb per day; too much food at this stage has been shown to reduce litter sizes, although very recently I have seen some contradicting research. Within our herd, very few sows can remain on 1 lb per piglet lactation ration right through to weaning even with a moderate-sized litter, and one once reached 3.5 lb per piglet reared by the time they were three weeks of age. In those circumstances, particularly if she has over 10 piglets, it can prove too much for the sow to eat in a day. With a particular sow we sometimes switch to a higher-calorie growers ration to input the required energy in a smaller volume as she won't eat enough sow rolls to keep her condition, but if you do this you have to be careful to choose one with similar copper levels to the sow diet, or copper poisoning is a possibility.

Other breeders in contrast have very different experiences: a Berkshire breeder told me that one lactating sow that raised a litter of 13 for a full eight weeks, was fed a maximum of 9 lb per day and maintained her body condition beautifully. Modern breeds are

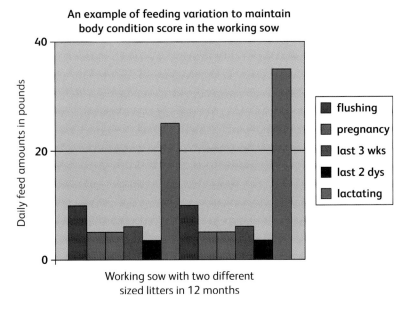

FIGURE 9.1 A typical feed variation in the working sow to maintain a body condition score of 3 across the year.

often fed more, with 6.5 lb maintenance ration being quite normal, and correspondingly higher lactating rations. Amusingly, a Kunekune breeder joked that her sows have three crumbs throughout pregnancy increased to four crumbs in the third week of lactation if they ever start to lose any weight at all!

So while telling you how much to feed your working sows is impossible, informing you how to judge how much to feed your sows by body condition scoring should see you producing robust litters of piglets from healthy sows, no matter what breed you have chosen.

THE BREEDING CYCLE – SEASONS

A sow comes in to season or 'brims' every 21 days and is fertile for approximately three days, with her fertility peaking at around 24–36 hours. Signs to look for include an increased body temperature to the touch, or by thermometer if you wish, and the vulva beginning to swell, and in pale colours you can see a reddening. I am also convinced that they smell more 'piggy' during this time. If you apply a slight pressure to her back near her hindquarters, she will stand still and is almost immovable. This is known as the 'standing response' and is a good sign she is ready for mating. This standing response can also be seen when she is in the pen with other sows, as they may mount her and she will stand stock still. Also if she is the one doing the mounting, it is a sign that she is in season. Of course these are only indications and you need to check for them at least twice a day to increase your accuracy in determining peak times for conception.

GETTING YOUR SOW OR GILT PREGNANT NATURALLY

So you have a well-grown gilt and she is approaching the age where she could safely become pregnant, or you have bought an established sow who knows what she is doing. Now what? You have the choice between a natural mating and using artificial insemination. In both cases it is wise to select a boar that is not related to your pig. You may have heard of line breeding (mentioned earlier in this chapter – using related boars to enhance desired characteristics), but this a game for the advanced breeder as the difference between inbreeding undesired characteristics and successful line breeding is a fine one. Good health, vitality and strong immune systems are inherited characteristics, and while a 'one off' cross back to a parent or sibling is unlikely to cause an issue, repeated line breeding without thought is likely to introduce some undesired, often unseen, conditions.

Keeping a stud boar is not to be contemplated for the complete novice. Give yourself a couple of years' experience of keeping pigs, as they really are very different to handle and another level of respect is required. We were very lucky with our first pedigree boar (pet name Swaysie). We purchased him from the show circuit, so he was used to being handled and was a very relaxed chap indeed. He loved to come out and go for a walk and was very gentle with both us and his 'wives'. Our second boar, which we bred ourselves, was a different ball game: he was very gentle with his wives and also with us but he had a streak in him that could terrify a novice, and very firm handling was required at times when you wanted him to do something that he didn't want to do. He had on more than one occasion attacked the board and so you couldn't relax at all with him when he was out and about. To be honest, you shouldn't relax around any working boar and should learn to use one eye independently of the other to keep a close vigil. If I have to go into a boar pen to do anything that requires concentration or both hands in use, e.g. pregnancy scan a wife, top up bedding, I always wait for capable back-up.

The boar must have enough work to keep him happy and also have company for most/all of the time. In the wild, boars lead solitary lives and only meet up with sows for mating, but we have found that if they are on their own for too long between sows they can become over-excited as the sow enters the pen and they could really hurt them. To maintain maximum fertility, the more the boar ejaculates the better; they do 'pleasure themselves' between wives sometimes and that counts too, but they prefer to have wives available. All our boars always have company, even if it is a sow or two that they have already made pregnant; however, a lot of breeders only put the two together until the deed is done and then remove until the next day and repeat the process. There is an advantage to the latter method as the sow is not removed from the main group for any length of time and so squabbles do not occur when you put her back in. For us it would take up too much time and we are not always about when the sow is ready and I like the idea that all our pigs have some company, including the boar. If you choose our method, then she must be at least 28 days pregnant, preferably a bit longer, before being reintroduced to the main herd, to minimise the chance of her losing her piglets through stress. We also pen unfamiliar sows side-by-side for a week or two to minimise the squabbles. Interestingly, when we mix sows or gilts that are not pen mates, or that haven't been for a while, with the boar present, they hardly ever squabble, and the more senior the boar, the less this happens. These friendships are subsequently maintained when the females are removed and penned together.

If you want to breed using natural mating before you have enough sows to warrant the expense of a boar, or you feel that you need more experience before owning one, some breeders loan out their boars for a stud fee and you provide the boar's bed and breakfast needs. To register the birth of any piglets, you only need to own the registered, pedigree female (in both the UK and the USA), so any resulting piglets can still be pedigree even if the boar is also registered in another owner's name. You will need to keep the boar for at least six weeks so your sow has a minimum of two ovulation cycles when with her new 'husband'. Other breeders offer to take in your sow; again it should be for six weeks as a minimum and you pay a stud fee plus B&B for your sow.

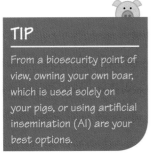

TIP

From a biosecurity point of view, owning your own boar, which is used solely on your pigs, or using artificial insemination (AI) are your best options.

You will need to pre-organise and book the boar with the breeder, and if the breeder also shows or exhibits pigs, then certain dates may be unavailable. You must also bear in mind that if you wish to sell your weaners, they will be ready to go eight to ten weeks after farrowing, so try to avoid dates when people don't buy eight-week-old weaners, such as in the middle of a thick snowy winter.

It is unlikely that smallholders who have a higher herd-health status will allow their boar to be taken off-farm or allow other sows on site. In the past we have taken in sows from people we knew well and whose pigs were well kept, but we have now decided that we will no longer practise boar hire and I suspect it will become increasingly harder to hire the use of a boar.

TIP

Your sow may not become pregnant the first time, so make sure you can be around three weeks after the initial dates as well if your sow isn't in the same pen as the boar.

Boar Behaviour

If the gilt/sow is running with the boar, he will start to hassle her a couple of days before she is ready, nudging her under the hind legs, foaming at the mouth, performing a sexy dance with his front legs with steam-train-like grunting noises, and exuding a sexy boar

FIGURE 9.2 *The mating of a sow and a boar.*

smell that can make your eyes water sometimes. He will try to mount her, but if she is not ready she will just walk away, and if she is clever, straight into the ark to efficiently knock him off.

Once mounted, the boar's corkscrew-shaped penis will become erect and enter the vagina, and the pair will remain locked together for up to 10–15 minutes (Figure 9.2). This prolonged mounting can be problematic if the boar is very large compared to the sow and there is a possibility she could collapse or weaken her hind legs. This is more common in young gilts that are being covered for the first time, so pick a similar-sized boar.

Young boars can be clumsy lovers, throwing themselves at the gilts at most opportunities, but if their sizes are reasonably matched or the boar is slightly smaller, then no harm should be caused. Some boars are rapists and so care should be taken if you have one of these; especially make sure young gilts aren't scared off by the whole experience and try leaving dominant pregnant sows in with these boars for company to help ease their vigour.

We have found that even small boars can mate with larger sows or gilts and it seems that 'where there's a willy there's a way', but don't scare them with a big dominant sow if they are less confident or allow them to strain themselves in the process! The boar may cover the sow a few times during the peak of her season but, as you may miss the actual covering, it is worth noting down the date the sow was first put in with the boar, to calculate the first possible date she could farrow. Also note the days you see any signs of mating, to calculate the probable farrowing date (see the born/covered/wean table, 9.1). If it's muddy, then look for scrape marks down either side of the sow, and also look for foam and dribble on her back. After intercourse there is sometimes a white jelly-like protein plug in the entrance to the sow's vulva or on the floor – if the boar hasn't eaten it (Figure 9.3). There may also be evidence of semen around the vulva. Boar sperm are not good swimmers, and they have quite a long way to travel, so to compensate, nature has given boars between 50–450 ml of ejaculate, which when compared to the few millilitres produced by other mammals, including humans, is quite something.

FIGURE 9.3 A protein plug that is deposited after the semen has been ejaculated – the brown is attached mud because it was retrieved from the pen.

Wait three weeks then look for signs of the sow being in season again. If she shows no signs, it is likely she is pregnant and you can work out her farrowing date with reasonable accuracy. If he covers her again, then she did not conceive during the last cycle and the whole waiting process begins again. The boar is likely to start becoming interested again before the sow or gilt is willing to comply and that gives you an early warning that she didn't take on the last cycle.

We have had a couple of cases over the years where the sow was definitely pregnant but then did not give birth, so you need to be aware that sows can reabsorb their piglets and come back into season. It is more common at certain times, e.g. in hot weather or shortening daylight. In combination with not conceiving at all, this is known as seasonal infertility and worthy of knowing about. Sows may also reabsorb if there are less than four piglets conceived, as there are insufficient hormones present to sustain the pregnancy – however, I have heard and experienced so many exceptions to this rule that I am beginning to doubt if it's even true, at least in traditional breeds. We have been caught out twice with pregnant sows returning into season and so always now ultrasound scan at 25 days and 50 days.

Putting Her Back in with Previous Friends After Being in with a Boar

With AI you can put her back in with her friends straight away without a problem. If you choose to let the boar cover her and then put her back in immediately or after a couple of days, then again no problem. If, however, you keep the sow or gilt in with the boar for three to six weeks or longer, you will need to take care when reintroducing her to the group. Mixing the sows or gilts back in with even previous friends can induce a fight, so if your sow has been out of the group for a while, then wait until she is at least 28 days pregnant before mixing. As I have said before, we always put them in adjacent pens for a week or two before mixing and feed either side of a fence line close together, which helps lessen the intensity and duration of the squabble. Before you then introduce them to the same pen make sure the pen is big enough that the non-dominant sow can run away, and put an extra water container on the far side to the main one. Try and mix nearer bedtime with a big feed for all and put a few bales of straw around so they can use them as barriers if required.

GETTING YOUR SOW OR GILT PREGNANT USING ARTIFICIAL INSEMINATION

There are numerous advantages to using artificial insemination (AI):

- No annual costs for a boar's upkeep.
- You can use different boar lines and improve the quality of your breeding and your herd's genetic diversity.

- In the UK, you don't get put on a 20-day standstill (see Chapter Two).
- There is no worrying about the wrong sows escaping and getting in with the boar.
- You don't have a boar to get rid of when he reaches the end of his working life.
- You won't have a bored and frustrated boar short of work.
- There is no problem trying to match the size of boars and gilts.
- Biosecurity isn't compromised by borrowing a boar.

The downside is that AI is not as reliable at getting your gilt/sow pregnant. Boar sperm is very fragile, sensitive to extremes of temperature, has a shelf life of approximately five to six days and has to go through the postal system to get to you. Also, you have to accurately detect your gilt/sows oestrus, order the sperm in time and perform the technique well enough to ensure a pregnancy.

Before Ordering Your Semen

Most important: you will need to get to know your sow's ovulation cycle, also called her 'season' or 'heat'. You can do this by observing her behaviour, particularly the 'standing response' and visible signs, both mentioned earlier. Oestrus detectors are available, and it is also said that if you look at a sow's saliva under a light microscope then you will see a ferning effect – if you have a light microscope handy, that is. Keep a diary of her seasons for a couple of months before you wish to mate her and look for the signs every three weeks. If they occur again, it is likely you have determined her seasons correctly.

On the first day of her season, as determined by her standing response, order your specific breed/bloodline semen, which arrives by courier the next day ready for use. The boar semen you buy is from registered boars, and so you can still register the birth and have pedigree piglets – exactly the same as natural covering. There is a list of UK and USA suppliers in the Useful Contacts section.

AI is not difficult to perform, but don't rush. Instructions can be found on the semen suppliers' websites or may arrive with each batch of semen. The hardest part, as previously mentioned, is to accurately diagnose her season before ordering. To help you out a little, if your sow's seasons are not so obvious and you don't have a boar handy to titillate her, there is a pheromone 'boar smell in a can', e.g. Boarmate™, which you can use to effectively tune her in to the task in hand (Figure 9.4). It can make the standing response easier to spot with varying degrees of success. Maiden gilts are the most difficult to precisely detect, and most accuracy is achieved if the season is induced artificially with hormones, e.g. Regumate™ or PG 600. Hormone treatment is expensive and goes against the 'natural approach', but it is worth knowing that such products exist.

The easiest to determine are sows that have recently been weaned, as they reliably start to come back into season between four to five days after the piglets have been removed. It is said that the ideal is to wean on a Thursday, she will start to come into season on Sunday, order your semen on Monday and use it on the Tuesday afternoon and again on the Wednesday morning and afternoon. As a *Commuter Pig Keeper* you may wish to wean on a Sunday, order your semen on the Thursday and use on Friday afternoon and Saturday morning and afternoon; but you do run the risk of it not arriving on the Friday and sitting in the post office all weekend, especially if you are not in to sign for it, as it is usually special delivery! For this reason you rarely get semen deliveries on a Monday, so bear that in mind when you are weaning. Your package will be delivered in a chilled/temperature-regulated polystyrene box, by a delivery guy with one eyebrow raised if the package has a prominent label stating the contents.

FIGURE 9.4 A can of pheromone spray.

The AI Procedure

In the UK, the semen usually arrives as a kit complete with the disposable applicators and three containers, either small bottles or a flexible pack, with 50–100 ml of semen suspended in a special life-extending liquid in each. In the USA you may have to order the disposable applicators separately, but most suppliers of semen stock them. You will also have to order the number of doses you want – order at least two.

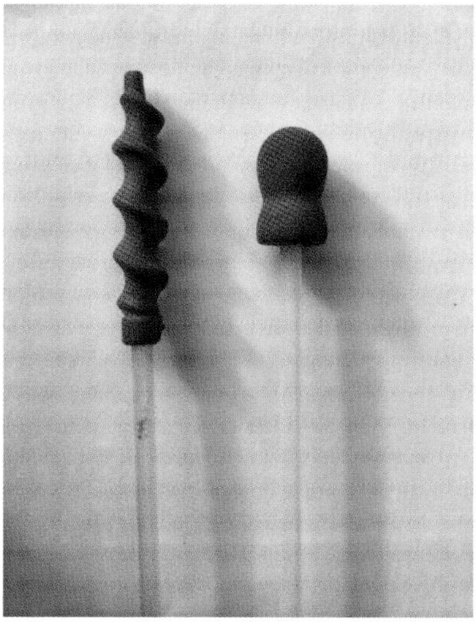

Don't put the semen in the fridge but place the box in a cool place. The semen lasts for approximately five to six days from collection, so it is important to order it as you require it. The applicators you are likely to get are the screw-in types (Figure 9.5), but there are others on the market that can be supplied, such as foam plug-in +/– lubricant and foam plug-ins with an internal, extendable tube – also known as 'squeeze and please' catheters. You can get ones especially for gilts, which have a smaller tip to insert into the cervix. I prefer the foam plug-in types as they are a bit softer.

FIGURE 9.5 Screw-in and foam-top insemination straws.

Preparation

Before approaching the pig, you will need one of the semen containers, one applicator, scissors to cut the semen bottle top (if present), damp tissues to clean the sow's vulva of mud, Boarmate™ if you are using it, an old tea towel, KY Jelly or similar non-spermicidal lubricant, a bucket of food and a helper. While you are getting ready, pop the semen container next to your skin for ten minutes to gently warm it up and very gently invert it to resuspend the semen from the bottom of the container. Don't shake it like a Formula 1 winner with a champagne bottle, as the sperm are very fragile. The gentler you are, the more intact sperm you will have – so treat it like a sleeping baby.

Insemination

If your helper is tall enough, get them to sit astride the sow facing her tail, which they can then hold out of your way, or ask them push down reasonably heavily on the top of her bottom.

1. Position the sow near a boar. He will start to perform a sexy dance, steam-train-like grunts and emit a powerful smell to get her in the mood. If you don't have a boar, liberally spray your Boarmate™ around her head, or onto dry tissue or towel and waft it around her snout.
2. Wipe the sow's vulva with your damp tissue to clean it; you don't want to push in foreign bodies, so just the outside is all that's required. Do not use soap or you will kill the sperm.
3. Lubricate the end of the applicator and insert it, pushing it upwards towards the spine and gently in along the inside wall of the vagina; this is to avoid accidental insertion into the bladder. Slowly straighten the applicator, and once you hit resistance you have reached the cervix. The older/longer the sow, the more of the applicator will disappear.
4. If using the screw-type applicator, rotate it anti-clockwise until it will not easily turn any more – which is around four turns. The sow's cervix has the corresponding screw thread to accommodate the boar's penis, which is spiral: nature is very clever. If using a foam plug top, just push it in gently until you feel some resistance. Very gently pull back on the applicator – if it doesn't immediately move backwards it is likely you are in. Don't yank on it though, as even if it is in place it might still come out and possibly hurt the sow. Now attach the semen container and hold it up higher than the sow – gravity is your friend here. Get your helper to push the sow's belly sides about. This assists in the release of eggs; the boar does it as well before he gets on board.
5. Give the semen container a little initial squeeze. If it starts to come back out over your boots, it is not correctly positioned and so try again. If your applicator is

spiral or plug top, then you must wait for the sow to suck the semen in with her own reflexes. This can take quite a long time, so spray a bit more Boarmate™ about and ask your helper to sit down firmly on the sow's back to represent the weight of the boar. When it does get sucked in, you may have to release a bottle-type container once or twice from the applicator to allow air back in, but do it quickly so you don't miss the moment! Don't squeeze the semen in (although tempting), as boar sperm are notoriously bad swimmers and so they will be exhausted before they reach the eggs and you will have wasted your money and effort. If your applicator is plug top with an extendable tube, you may gently squeeze the semen in as the internal tube reaches to the far end of the cervix, dispensing the semen into prime position (Figure 9.6).

6. Once finished, give your sow a few moments before unscrewing the applicator clockwise (or pulling if it was a plug-in type), and give her a bit of feed if she is interested. As mentioned earlier, when a boar ejaculates, he deposits a protein plug in the exit to keep the sperm contained. AI doesn't do that, so it's a good idea to keep the sow still for a while.

7. Repeat the process for the remaining containers at specified intervals. If you caught the season early, then perform 12 hours apart, but if you think she is close to the

FIGURE 9.6 A sow being inseminated.

end, perform the three procedures eight hours apart. Her fertility peaks at approximately 24 to 36 hours and so you want at least two AI applications to be within that time. If the third bottle just doesn't get taken up by the gilt or sow, then we have found that's a good sign that the previous two bottles have done the trick.

8. Put her back in her pen and wait three weeks to look for signs again, and possibly you will have to repeat the procedure – keep your fingers crossed. You can of course confirm with a pregnancy ultrasound scanner from day 18 post-insemination (if skilled) and reliably from day 25 but you may miss ordering the next batch of semen in time if you only rely on the scanner.

DETECTING A PREGNANCY

Visual detection isn't always reliable. Some sows show they are pregnant very early on, especially if they have had previous litters: their underlines drop and their sides expand and they can't wait to let it all go. Others, especially first timers, may not show until a few weeks before and you can be scratching your head wondering 'Are they? Aren't they?' until near the due date, especially if they weren't ultrasound scanned.

- Check the vulva daily for mucoid tackiness from days 14–21. I suspect this isn't very reliable as I haven't noticed any discharged mucus during a pregnancy.
- Visually check the sow for enlargement of abdomen, teats and mammary vein from day 80 of pregnancy.
- Boars can detect an ovulating pig before most modern equipment, and if the boar isn't interested and the female isn't exhibiting the standing response, then it's a good sign she is 'in pig'.
- If you don't have a boar to wander your sow past, then there are relatively inexpensive ultrasound scanners available (2016: £320/$550) in both the UK and USA (Figure 9.7). Use on days 21–25 and between days 45–90 to obtain a reliable indication of a continuing pregnancy.
- Of course you can hire livestock services or your vet to come in and do it, who may have a scanner with a screen to tell you how many piglets as well.

There are highly accurate invasive scanning devices, serological testing and vaginal biopsies that can be performed, but I wouldn't want to go down those routes.

There are also a couple of 'alternative methods' that could be fun to see if they work. If you know the look of the lower part of your gilt or sow's vulva, then you may see a difference in the direction their clitoral hood (located at the bottom of the vulva) points when pregnant. In non-pregnant gilts, it usually points down and in the experienced sow it either points down or out at a horizontal angle. As the pregnancy progresses, the

FIGURE 9.7 A sow being ultrasound scanned for a pregnancy.

Photo courtesy of Liz Shankland.

weight of the expanding uterus drags everything down, causing the clitoral hood to point up. With larger litters this effect may be noticeable within a few weeks. What you are looking for is a change in direction. I have also been reliably told, but never tried it, that human pregnancy dipstick tests you can buy at supermarkets and drug stores also work on pigs. I would have thought they wouldn't detect a pig hormone, but perhaps they are similar enough in the biochemical make-up to make it possible. I suspect a high level of false negatives would occur rather than false positives.

PREGNANCY

Sows are pregnant for three months, three weeks and three days (average 115 days; normal range 111–120 days), but like all animals this can vary slightly. Ideally, keep her in a familiar routine, including her normal feed ration, to minimise stress and increase the implant rate of the fertile eggs. The viable foetuses secrete the pregnancy mainte-nance hormone progesterone, informing the sow she is pregnant. Variation in pregnancy length can be due to environmental conditions, specific breed, or time of the year. It's also recognised that gilts tend to have a shorter pregnancy length, and in larger litter sizes pregnancy also tends to be shorter than in smaller litters – possibly due to lower levels of corticosteroids being collectively secreted.

Newly pregnant sows should not be mixed with new pigs without gentle introduction. She can remain in her existing group right through to farrowing, but the other pigs are likely to show a lot of interest and increase the chances of her piglets being squashed. So a week or so before her due date it is advisable to give the sow separate accommodation, which has been cleaned, disinfected and bedded down with straw.

I do have a friend who rears her small herd of five sows and one boar all together with great success. The piglets climb all over the boar, who likes to protect them, and they cross-suckle with the other sows, which is great for the immunity of the piglets. For us this would be a nightmare as we like to time our piglets for the show circuit and also we need to know exactly which piglet is from which sow so we can maintain our bloodlines accurately. We do let our some sows mix together, with separate arks provided, after the piglets have been identified by being notched or tattooed and let the piglets cross-suckle, but you have to know the sows well and be sure that they will not fight.

Just before moving to the farrowing quarters is also a good time to deworm her against parasites, and if you vaccinate your pigs, then a booster about three to four weeks before farrowing will ensure the transfer of high antibody levels to the milk. We always farrow outside in arks without heat lamps but some people prefer to farrow their pigs in a larger building as they have better access to the sow should she have trouble during the birth, and so they have plenty of headroom and probably electric lighting.

We have farrowed in a stable on a couple of occasions, and although it is easier during the birth, I find that before and after the birth it's a pain. Not only do you have to muck out every day, which takes more time, but you will need a heat lamp or heat source of some description in the colder months, as a sow's body heat can warm up an ark quite nicely but could not warm up a stable. The piglets also then require moving at weaning to another pen and I like to keep the routine as familiar to them as possible while they are adjusting to their non-milk diet. What I would advocate is the use of a pig ark with a rear door so that access to the farrowing pig is easier should it be required, or using a stable with direct access to some mud outside.

FARROWING (GIVING BIRTH)

The piglets are responsible for their actual birthdate by secreting corticosteroids stimulating the placenta/uterus to produce the hormone prostaglandin. This directly causes the piglets to cease secreting the pregnancy-maintenance hormone progesterone, initiating the birthing process. The normal farrowing process has a huge variation of birthing indicators and in the times taken. 'Normal' will vary from sow to sow, and even between different litters from the same sow, so use the listed signs as a guide to knowing what is acceptable rather than a prescription of what is going to happen.

The preparation for farrowing gradually begins approximately two weeks prior to the actual date, with teat enlargement and the increase of prominent veins in the udder due to an augmented blood supply to the area. The vulva begins to swell and may redden around four days before farrowing. The mammary glands become taut, triangular and more defined about two days before farrowing (Figure 9.8). Watery secretions may be seen two days prior, which become more 'milky' within 12–24 hours. If milk is abundant and easily flows when the teats are gently squeezed, then sow is probably within six hours of farrowing. Restlessness and nest-building are signs that the birth will be in the next eight to 24 hours, with sows weaving straw, grass, twigs and any other available material into the most impressive constructions. Their respiration rate may also increase as they prepare, peaking around nine to ten hours before farrowing. Finally the cervix dilates, opening up the exit route.

About an hour before giving birth, the sow's activity may calm and she may lie quietly on her side in her nest before the straining begins, although every sow is different. Some sows can be increasingly agitated and restless, as the uterine contractions intensify. If they are very restless, then a light sedative can be administered. Small amounts of red-tinged fluid may pass from the vulva, occasionally with pellets of meconium (faecal matter passed by the piglets before they are born), although I have never once seen this. Table 9.1 shows a timeline of a normal farrowing and what to expect.

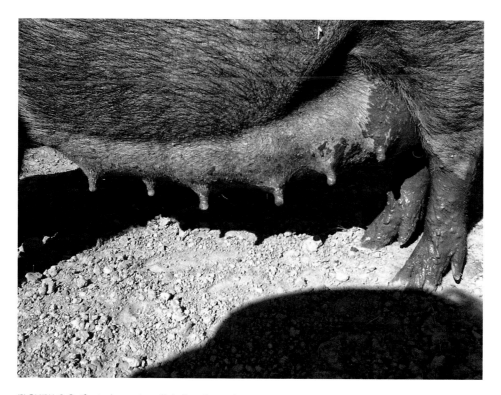

FIGURE 9.8 Teats heavy in milk before farrowing starts.

An intermittent abdominal muscle straining occurs before the birth of the first pig-let, usually accompanied by shivering, with the sow drawing her upper hind leg upwards. After the first-born, the straining usually becomes milder, except just before a piglet expulsion. Don't be surprised if there is a 45-minute delay between the first and second piglet but from then on 10–20 minutes between live piglets is normal (range: minutes to a couple of hours) and a slightly longer delay between stillborn piglets of 35 minutes plus, but that does fall between documented normal ranges.

The majority of piglets are born head first with the front legs folded back (anterior presentation) (Figure 9.9), but towards the end of farrowing there are more pigs pre-sented backwards with the rear legs first and the front legs extended under the chin (posterior presentation). The twitching of the sow's tail is a good sign that another piglet is due to arrive in the next ten seconds. The whole birthing process takes between three and six hours (range: one to ten hours).

Newborn piglets are still attached to their umbilical cord and it breaks as they strug-gle and try to walk; if the sow does not object, you can spray the ends of the cords with antiseptic or iodine. We never do this, because if the piglets squeal, the sow will get up and down protecting them and she may well squash the others while doing so. The umbilical cord is ridiculously long (circa 25 cm /11 inches) and trails behind the piglet as it walks; the piglet can get caught up in it, but leave it alone. I think it is nature's way of preventing infections (Figure 9.10); the end of the cord is essentially an opening into the piglet, and so if it is very long, any infection getting in has to travel a long way to the piglet. By that time, the cord has dried up and the bug has been stopped – Mother Nature is very clever.

FIGURE 9.9 A piglet being born.

FIGURE 9.10 The long umbilical cord.

FIGURE 9.11 The afterbirth.

Photo courtesy of Lorraine Jones of Le Logis France.

Experienced sows are the least likely to get up and down during the birth; that behaviour is usually confined to first-time gilts, and contrary to some books, the sows and gilts can deliver their piglets from both uterine horns without the need to get up or turn over. Sows may remain inactive for up to 95 % of the first 24–48 hours.

The placenta/afterbirth is typically expelled within four hours after the last piglet (range: zero minutes to over 12 hours) (Figure 9.11). If you are present, within safe reach and know your sow well, check that there isn't another piglet caught up in the afterbirth, as occasionally it will be passed during the farrowing process.

Sometimes the afterbirth is delivered in two halves, one from each uterine horn, and sometimes whole at the end. A retained placenta is rare, and the failure to pass the afterbirth is often indicative that there is another piglet or piglets remaining in the birth canal.

Once the afterbirth has been passed, the sow should appear contented. She should be continually talking to her piglets, and the shivering and top hind leg movements will cease. There may be a slight to heavy discharge for up to five days; provided the udder is normal and the sow is eating well, then it's a natural post-farrowing process. If you have had to carry out an internal examination or assist in delivering piglets, give antibiotic cover to guard against infection.

Some sows will eat their afterbirth and any dead piglets and others won't. Those not eaten must be disposed of in a licensed incinerator or by some other approved method.

When to Assist Sows During Farrowing

Signs of dystocia or 'farrowing difficulty' include anorexia; prolonged pregnancy length; bloody or foul-smelling discharge; piglet meconium passed without straining; prolonged labour; straining without piglets appearing; sow exhaustion and cessation of labour; and a distressed sow. The most common cause, however, is a piglet or two positioned incorrectly and blocking the birth canal.

Manual examination of the vagina and cervix to remove an obstruction or to correct the presentation of a piglet ideally requires strict sanitation using hot water and soap or obstetric gel, the use of obstetrical gloves, sterile lubricants applied liberally and the most gentle of approaches. If you are not competent at performing manual piglet extractions, then don't 'have a go' unless you absolutely have no other option, and bear in mind to be as gentle as is humanly possible or you could cause a serious welfare issue.

If the sow is lying on her left side, use your left hand to explore the birth canal; if she is lying on her right side, use your right hand. Begin by placing two fingers very gently inside the vulva, checking for piglets. If you can't feel anything, shape your hand and fingers like a cone and very gently push your lubricated hand into the vagina. In a large, well-dilated sow, the vagina, cervix and uterus tend to blend together. Generally, it's not necessary to reach past your elbow. Stop once you reach a piglet. If it's coming head first, wrap your thumb and forefingers around its ears and jaw and pull gently. If the birth canal is too tight, grasp the lower jaw with your thumb under the tongue and your index finger in the V-shaped bones of the lower jaw. Or, place your thumb over the piglet's snout and your index finger snugly behind the upper needle teeth for a firm grip. Squeeze and pull gently. If the pig is presented backwards, place your index finger between the legs, place your thumb on the outside of one leg and your middle finger on the outside of the other leg, squeeze together and pull gently. When the piglet is out, do not break the umbilical cord, so the blood supply continues until it breaks naturally, but do remove the placental tissues from around the piglet's snout.

Don't go to any extraordinary efforts to reach in for more piglets, particularly if you have released a blockage; assess each piglet's delivery needs individually. Most often the sow will deliver the remaining piglets without assistance. The manual extraction of piglets is by far the safest technique. The use of forceps, cable snares or blunt hooks risk trauma to the sow's reproductive tract and to the unborn piglets and should be strictly for professional use.

Some Abnormal Presentations

- Breech presentation is where the piglet is presented backwards but with the hocks and legs tucked upwards causing the rump to get caught on the brim of the pelvis. This can be remedied by gently sliding your hand through birth canal until you touch the pig and hooking your finger around the pig's hock, then pulling gently toward you, straightening the legs out behind it. This will put the pig in a normal, posterior presentation, making it easy to pull the pig gently out.

Can You Breed Pigs and Work Full-time?

- The uterus has two horns and one exit, and so it's no surprise that occasionally two pigs are presented for birth at the same time, effectively blocking the exit. When you do go inside to sort it out, one may be coming backwards, the other forwards, or both may be facing the same direction. If the uterus is large enough to pass a hand through, deliver the first pig you reach, and then return for the second.

- A heavy uterus is where the weight of a large litter has dragged the uterus down along the abdominal wall, causing the birth canal to form an 'S' shape. The uterus sometimes cannot contract enough to push the pigs up and over the pelvic rim, so assisting in the delivery of the first piglet may be required. This often smooths out the S shape and the remainder of the piglets are born normally.

- Oversized piglets (over 1.8 kg/4 lb) in an under-developed (young gilt) or swollen birth canal may become lodged or just plain exhaust the sow. This can be a difficult problem to correct, but find the smallest experienced hands you can and use plenty of lubrication. A recommended method is to use a 2 m piece of disinfected nylon cord, passing the cord 5 cm behind the left and right ears and twisting lightly under the jaw to secure before pulling gently in a downward movement to bring the piglet out. It's recommended that you familiarise yourself with the technique by cutting off the end of a wellington boot, placing a dead piglet inside with its head presented to you and practising placing the cord around the neck.

- Uterine inertia, sluggishness or a lack of contractions are among the most common complications at farrowing time. The list of causative factors is long but includes the use of prostaglandins to induce farrowing or the use/over-use of oxytocin (see later in this chapter) to induce uterine contractions. You can normally avoid the use of oxytocin by inserting a clean arm gently into the vagina, which is known as 'sleeving' or 'feathering'. Piglets suckling the sow's teats also stimulate uterine contractions, so massaging the udder and teats may be helpful.

- Hypocalcaemia – a deficiency of calcium – is another common disorder in older sows, i.e. those that have had three or more litters. If you 'sleeve' the sow and there is no uterine tone (you cannot feel any muscles contracting) and you are confident there are no stuck piglets, then you can administer oxytocin and check her again. If there is still no tone, an intramuscular dose of calcium, as recommended by your vet, may return normal uterine tone within 15–20 minutes and she will continue farrowing normally.

- Mummified babies are usually passed through the cervix without issue as they are comparatively smaller than live piglets, but they don't stimulate uterine contractions and if too slow exiting, they may be followed by a stillborn piglet.

- If a piglet is delivered and it fails to breathe, take a small piece of straw and poke it up a nostril. This usually invokes sneezing or coughing reflexes, effectively removing any mucus blocking the windpipe. If this doesn't work, place your middle finger across the mouth of the piglet with its tongue pulled forward, cradling the

TABLE 9.1 Gestation chart showing date of mating, approximate date of birth and date of weaning.

	1	2	3	4	5	6	7	8	9	10	11	12	13	14	15	16	17	18	19	20	21	22	23	24	25	26	27	28	29	30	31	
Bred January	1	2	3	4	5	6	7	8	9	10	11	12	13	14	15	16	17	18	19	20	21	22	23	24	25	26	27	28	29	30	31	Bred January
Due April	25	26	27	28	29	30	1	2	3	4	5	6	7	8	9	10	11	12	13	14	15	16	17	18	19	20	21	22	23	24	25	Due May
8weeks JUNE	20	21	22	23	24	25	26	27	28	29	30	1	2	3	4	5	6	7	8	9	10	11	12	13	14	15	16	17	18	19	20	8 weeks JULY
Bred February	1	2	3	4	5	6	7	8	9	10	11	12	13	14	15	16	17	18	19	20	21	22	23	24	25	26	27	28	–	–	–	Bred February
Due May	26	27	28	29	30	31	1	2	3	4	5	6	7	8	9	10	11	12	13	14	15	16	17	18	19	20	21	22	–	–	–	Due June
8weeks JULY	21	22	23	24	25	26	27	28	29	30	31	1	2	3	4	5	6	7	8	9	10	11	12	13	14	15	16	17				8 weeks AUG
Bred March	1	2	3	4	5	6	7	8	9	10	11	12	13	14	15	16	17	18	19	20	21	22	23	24	25	26	27	28	29	30	31	Bred March
Due June	23	24	25	26	27	28	29	30	1	2	3	4	5	6	7	8	9	10	11	12	13	14	15	16	17	18	19	20	21	22	23	Due July
8weeks AUG	18	19	20	21	22	23	24	25	26	27	28	29	30	31	1	2	3	4	5	6	7	8	9	10	11	12	13	14	15	16	17	8 weeks–SEPT
Bred April	1	2	3	4	5	6	7	8	9	10	11	12	13	14	15	16	17	18	19	20	21	22	23	24	25	26	27	28	29	30	–	Bred April
Due July	24	25	26	27	28	29	30	31	1	2	3	4	5	6	7	8	9	10	11	12	13	14	15	16	17	18	19	20	21	22	–	Due August
8weeks SEPT	18	19	20	21	22	23	24	25	26	27	28	29	30	1	2	3	4	5	6	7	8	9	10	11	12	13	14	15	16	17	–	8 weeks OCT
Bred May	1	2	3	4	5	6	7	8	9	10	11	12	13	14	15	16	17	18	19	20	21	22	23	24	25	26	27	28	29	30	31	Bred May
Due August	23	24	25	26	27	28	29	30	31	1	2	3	4	5	6	7	8	9	10	11	12	13	14	15	16	17	18	19	20	21	22	Due September
8weeks OCT	18	19	20	21	22	23	24	25	26	27	28	29	30	31	1	2	3	4	5	6	7	8	9	10	11	12	13	14	15	16	17	8 weeks NOV
Bred June	1	2	3	4	5	6	7	8	9	10	11	12	13	14	15	16	17	18	19	20	21	22	23	24	25	26	27	28	29	30	–	Bred June
Due September	23	24	25	26	27	28	29	30	1	2	3	4	5	6	7	8	9	10	11	12	13	14	15	16	17	18	19	20	21	22	–	Due October
8weeks NOV	18	19	20	21	22	23	24	25	26	27	28	29	30	1	2	3	4	5	6	7	8	9	10	11	12	13	14	15	16	17	–	8 weeks DEC
Bred July	1	2	3	4	5	6	7	8	9	10	11	12	13	14	15	16	17	18	19	20	21	22	23	24	25	26	27	28	29	30	31	Bred July
Due October	23	24	25	26	27	28	29	30	31	1	2	3	4	5	6	7	8	9	10	11	12	13	14	15	16	17	18	19	20	21	22	Due November
8weeks DEC	18	19	20	21	22	23	24	25	26	27	28	29	30	31	1	2	3	4	5	6	7	8	9	10	11	12	13	14	15	16	17	8 weeks JAN

Can You Breed Pigs and Work Full-time?

	1	2	3	4	5	6	7	8	9	10	11	12	13	14	15	16	17	18	19	20	21	22	23	24	25	26	27	28	29	30	31	
Bred August	1	2	3	4	5	6	7	8	9	10	11	12	13	14	15	16	17	18	19	20	21	22	23	24	25	26	27	28	29	30	31	Bred August
Due November	23	24	25	26	27	28	29	30	1	2	3	4	5	6	7	8	9	10	11	12	13	14	15	16	17	18	19	20	21	22	23	Due December
8weeks JAN	18	19	20	21	22	23	24	25	26	27	28	29	30	31	1	2	3	4	5	6	7	8	9	10	11	12	13	14	15	16	17	8 weeks FEB
Bred September	1	2	3	4	5	6	7	8	9	10	11	12	13	14	15	16	17	18	19	20	21	22	23	24	25	26	27	28	29	30	–	Bred September
Due December	24	25	26	27	28	29	30	31	1	2	3	4	5	6	7	8	9	10	11	12	13	14	15	16	17	18	19	20	21	22	–	Due January
8weeks FEB	18	19	20	21	22	23	24	25	26	27	28	1	2	3	4	5	6	7	8	9	10	11	12	13	14	15	16	17	18	19		8 weeks MAR
Bred October	1	2	3	4	5	6	7	8	9	10	11	12	13	14	15	16	17	18	19	20	21	22	23	24	25	26	27	28	29	30	31	Bred October
Due January	23	24	25	26	27	28	29	30	31	1	2	3	4	5	6	7	8	9	10	11	12	13	14	15	16	17	18	19	20	21	22	Due February
8weeks MAR	20	21	22	23	24	25	26	27	28	29	30	31	1	2	3	4	5	6	7	8	9	10	11	12	13	14	15	16	17	18	19	8 weeks APR
Bred November	1	2	3	4	5	6	7	8	9	10	11	12	13	14	15	16	17	18	19	20	21	22	23	24	25	26	27	28	29	30	–	Bred November
Due February	23	24	25	26	27	28	29	30	31	1	2	3	4	5	6	7	8	9	10	11	12	13	14	15	16	17	18	19	20	21	–	Due March
8weeks APR	20	21	22	23	24	25	26	27	28	29	30	1	2	3	4	5	6	7	8	9	10	11	12	13	14	15	16	17	18	19		8 weeks MAY
Bred December	1	2	3	4	5	6	7	8	9	10	11	12	13	14	15	16	17	18	19	20	21	22	23	24	25	26	27	28	29	30	31	Bred December
Due March	26	27	28	29	30	31	1	2	3	4	5	6	7	8	9	10	11	12	13	14	15	16	17	18	19	20	21	22	23	24		Due April
8weeks MAY	21	22	23	24	25	26	27	28	29	30	31	1	2	3	4	5	6	7	8	9	10	11	12	13	14	15	16	17	18	19		8 weeks JUNE

head with the rest of that hand, hold the back legs with your other hand and swing the piglet firmly downwards to propel any mucus out. Be warned, I have been bitten more than once performing this and those needle teeth are blummin' sharp!

Use Oxytocin with Care

You shouldn't go around jabbing oxytocin into your sow 'willy-nilly' as it can have some devastating side-effects. It's a powerful prescription drug for use only when absolutely necessary and most definitely is not required routinely.

Before considering administering oxytocin, try to stimulate natural oxytocin release by vaginal palpation/sleeving, udder massage and, if possible, keeping at least four piglets nursing at a time. If you then decide you would like to administer some, first check that the cervix is fully dilated and that there isn't a piglet blocking the birth canal.

Use only if a minimum of 40 minutes have passed between births and limit the use to twice per sow. The preferred sites for administration are the neck or the vulva. Improper use of oxytocin has been shown to cause umbilical cords to rupture, leading to higher stillbirth rates.

High doses of oxytocin (over 20 IU) may additionally create a refractory period lasting up to three hours, in which the natural or injected oxytocin fails to stimulate uterine contractions – effectively reversing what you are trying to achieve.

Recent research has suggested additional negative effects on the foetus, including increased contraction of the heart muscle, decreasing the number of heartbeats, which has been shown to reduce oxygen levels in the piglets.

There is also an increased chance of the sow becoming hypocalcaemic due to force and frequency of the uterine contractions. Having said all that, oxytocin is a useful drug to have in the medicine box and the associated risks may outweigh the risk of not using it in some instances – just give it thought before administering.

Happy Days

Please be reassured that the normal farrowing scenario is the most likely outcome. Once the afterbirth is expelled and the piglets are all suckling, that's farrowing over. While farrowing problems can and do occur, make observation the most important tool, and try to leave them to it as much as possible. More problems are created by interfering than are solved.

TABLE 9.2 The signs and timeline of imminent farrowing.

SIGNS OF IMMINENT FARROWING	TIME BEFORE FARROWING	
	MINIMUM	MAXIMUM
Teat enlargement/development of udder	10 days	14 days
Swelling/reddening of vulva	4 days	6 days
Mammary glands become taut and triangular, may have watery secretions	2 days	4 days
Milky secretions	6 hours	1 day
Nest building/increased respiration	0.5 days	1 day
Lull in sow activity	1 hour	2 hours
Small amounts of red-tinged fluid	15 minutes	30 minutes
Abdominal muscle straining and shivering	Minutes	
Tail waggling	Seconds	

Immediately After Farrowing

After being born and suckling successfully, the piglets should fall asleep at the teats while the sow grunts to them softly. If the sow hasn't eaten the afterbirth, this and any dead piglets need to be cleared up when it is safe to do so, quicker in the warmer months or if you use a heat lamp. You can freeze the afterbirth and any dead piglets until you have enough to warrant a trip to an incinerator or to call in someone to collect them, but I would have a separate freezer dedicated to the purpose. The container they are transported in must also be sealable and leak proof.

Sows may not be interested in their food after giving birth for a short while, but their appetite should return within 24 hours. Keep a close eye on the sow for the first few days, as if she continues to be off her food and lethargic she could lose her milk supply due to mastitis or farrowing fever. If this happens call your vet or administer antibiotics immediately. If all is well, then your sow should be fed to maintain BCS 3 to 3.5 (see Chapter Three).

One problem that you hopefully won't come across on your first farrowing is when a completely novice sow seems calm and controlled throughout the birthing process until one of the piglets moves up to her head, and then she becomes frightened by the 'strange creature' she can now see.

She may even try to bite or kill the piglets when she gets up and sees them all around her. You or your veterinary surgeon can give her a dose of a sedative, e.g. Stresnil™, and she will sleep for a good few hours, giving the piglets a chance to feed and get stronger, but be careful if you are farrowing any distance from your house, as a cheeky fox or other predator may, if hungry enough, pop in and take a couple.

If you can collect the piglets safely, without having to sedate the sow, e.g. they are near the rear door of the ark, scoop them up and put them together until she calms down. It is rarely a permanent problem. The piglets will be fine for a while without feeding but it may be wise to have a heat lamp on standby.

A slightly more common problem in first-time gilts comes from trying to be a good mother, and when they go outside to eat, drink or go to the toilet, they suddenly remember they have babies and go rushing back in too fast and can kill piglets in the process. This usually calms down after the first day and is rarely a problem in subsequent litters. We have had two sows do this and both were fine with their second litters.

SOW AND LITTER CARE AFTER THE BIRTH

During the birthing process and the following 24 hours is when the sow is most likely to squash her piglets. When the sow stands up, the piglets congregate around her legs and can easily get trapped when she lies down again. You will hold your breath more than once when watching them. They learn quite quickly to move out of her way. The less overweight the sow is, the quicker she can get up when they squeal from underneath her. Keeping our sows at BCS 3 for the birth has saved many piglets from death.

The use of a farrowing rail or pen could help minimise squashed piglets, especially if you hang a heat lamp over the area, and they are out of the sow's reach. We had little success with farrowing rails, as our pigs seem to remove them with ease and throw them to one side. Perhaps this is more indicative of the quality of our rails.

We now farrow outdoors all year round in ordinary arks and do not use a heat lamp; the piglets keep warm by cuddling into Mum in a nice straw bed and by lying in a big heap; the sows even lie across the doorway to make it extra warm and cosy. I try at all costs not to use a heat lamp. In my opinion they just create the perfect environment for bugs to grow, and when piglets lie in a heap, they become highly visible to the sow and are less likely to get squashed. This does contradict most commercial pig-keeping advice that I have read, but I have only reared traditional or non-modern breeds and have seen only positive effects. If you have chosen a modern breed, then perhaps exercise some caution. I also accept this may not be possible in countries outside of the UK that have extremes of temperature.

By two weeks old, piglets are more capable of regulating their own body temperature by shivering, running around, lying in a heap or lying on cool earth. There are various optimum temperatures quoted for pigs and piglets, even in traditional pig-keeping books, which is pointless really as you can neither turn the sun on and off nor prevent snow! What you can do once they are a few days old is make sure there is plenty of dry bedding, shady areas and wallows (shallow in pens with piglets), and provide water in a vessel they can safely reach and get out off should they decide to climb in . . . which they will. In this way the pigs and piglets can choose their own optimum temperature.

Piglets feed little and often. At the beginning, the sow will lie down and softly grunt at the piglets to come and feed. As the piglets get older they hassle her with squeals and excited noises until she relents and lies down to let them feed. You will find that from around three weeks the sow will lie on her teats and not let them feed until she is ready, encouraging the piglets to find alternative feed from an early age. We introduce small pig grower pellets at between seven and ten days, as the piglets will start to eat very small amounts from an early age. We sprinkle them on the turfs of earth to encourage them to tuck in. Careful observation is carried out to ensure that no piglet is suffering from anaemia (see later in this chapter), especially if any refuse to leave the ark within a week. Always protect yourself from the sow or gilt when handling the piglets, especially if what you are doing will make them squeal, by removing her temporarily or blocking her way effectively.

As the piglets get older and are confidently eating their growers pellets, we slowly reduce the growers until at eight weeks they are eating sow rolls. After weaning they are given 2 lb of sow rolls per day each.

WEANING

Commercial piglets are often weaned at four weeks in the UK and from two weeks (usually three) in the USA; this is because the sow reliably comes back into season within a few days of weaning and commercial units have specialised housing to take the place of the sow in terms of warmth and milk-containing feeds. With the traditional and heritage breeds it is common to wean between six and eight weeks – sometimes ten. Try to leave the piglets with the sow for the full eight weeks, but if she has had enough and starts to toss her piglets all over the place, it is safer to wean them. We wean between six to seven weeks now, but keep the piglets until they are at least eight to nine weeks before selling them on, or moving them to another pen if we are keeping them.

Ideally, remove the sow from the pen and leave the piglets behind. It is unlikely she will even look back. An exception to this is our friend's Mangalitza herd: the sows pace vigorously, calling to their piglets, and the piglets whimper and cry for 24 hours. When our friend weans her piglets, she has to do it with them all out of earshot of each other.

How you feed the sow depends on whether she will be put straight back to the boar, have AI or be rested for a while. If she didn't maintain her body condition score of 3, then re-evaluate your feeding regime when your sows are lactating. If she is too thin but *not* going back to the boar, cut back by 1 lb per day until her teats start to dry up and then increase her feed accordingly to get her back up to condition.

Always keep your weaned piglets for two weeks before selling them on, to make sure they are all eating and drinking well and have no other problems that show under the stress of being weaned. The most stressful time immunologically and mentally for a piglet is just after weaning, so don't add moving home into the scenario just yet.

SPECIFIC HEALTH CONSIDERATIONS FOR BREEDING PIGS AND PIGLETS

None of the information here replaces qualified veterinary advice, and is just brief guidance on a few of the more common or avoidable problems. I strongly advise you to buy at least one comprehensive veterinary manual specifically for pigs.

Causes of Infertility – Sow

It is highly relevant to point out that, historically, pigs are seasonal breeders producing litters in the spring and not in the depths of winter. It is only with modern warmer husbandry methods and access to plenty of food that they have become year-round breeders.

Even in well-managed herds, there may be some seasonal infertility or autumnal infertility as nature battles nurture, which should ease as the daylight length starts to increase again. There are other non-infectious causes, including being cold, mouldy or unbalanced nutrition from home-made diets, stress, lameness, parasite burdens and a reaction to a vaccine or medication.

There are also infectious causes from bacterial infections (*Brucella*, *Leptospira*, *Escherichia coli*, *Klebsiella*, *Streptococci*, *Pseudomonas*) or viral infections (Aujeszky's (pseudorabies), porcine respiratory and reproductive syndrome (PRRS), porcine parvo virus (PPV) and Influenza). Inadequately treated or undetected metritis, cystitis and nephritis could also be a cause.

Sometimes sows just do not want to have sex with a particular boar and it is worth trying another boar if you have one or AI if she really does object. Some people use a stall so the sow cannot get away. 'Rape stalls' are pretty horrific, from what I have been told, so avoid them. The sow is better off culled than being aggressively raped. It should also be noted that, in some breeds, some sows can become barren by not farrowing often enough – or, in the case of gilts, soon enough. The information on these phenomena is scant and anecdotal, as most research is performed on commercial herds with a 'two misses and you're out of the herd' policy.

Causes of Infertility – Boar

Often the boar's fertility is overlooked when low litter numbers are born, with the sow taking the lion's share of the blame – but it can be the boar. Sterility is easy to spot, as no piglets are born, but sub-fertility or poor sperm quality may not be noticed for quite a while, especially in small herds.

Judging a boar's fertility before he reaches eight months of age is a little unfair as all his man parts are still maturing up until that point, so if you have used a very young boar, then cut him some slack until he is older. Like the sows, infertility can be caused by infectious organisms and non-infectious environmental causes. Extremes of temperature even for one day or a brief illness causing a febrile (high internal heat) episode can render a boar temporarily infertile for up to a few weeks. Reluctance to work due to lameness, being terrified by an aggressive sow when inexperienced or suffering penile/scrotal damage must also be considered.

NOTE: For both sows and boars, 'Inbreeding Depression' resulting from high Co-efficient of Inbreeding percentages (see Chapter Eight) could significantly affect their reproductive performance, including infertility, and congenital defects in the progeny.

Farrowing Fever – Metritis

After the sow has farrowed, if she is lethargic and refuses to rise, eat, drink and has not defecated, call the vet immediately. She may have metritis or what is colloquially known as 'farrowing fever'. It is a bacterial infection of the uterus and without prompt treatment her milk will dry up and she is at high risk of dying. If you have antibiotics, administer some as soon as possible, as this will help fight infection and bring her temperature down, which will be raised if she has an infection. Put it in your medicine book as well. The sooner the sow starts eating and drinking, the higher the success of her raising her piglets. If the sow's milk has started to dry up, provide the piglets with an artificial milk source in a shallow tray or bowl until she is back up and running (hopefully).

Farrowing Fever – Mastitis

The sow not eating is one of the first signs of mastitis. It most commonly occurs a few days after giving birth but can also happen at any time during lactation or shortly after weaning. The sow's udder will be hot to touch and may feel lumpy or hard if the infection has established. It will be painful, so go easy on the prodding. Hungry piglets caused by a reluctance by the sow to let them suckle is another sign. At first they will be quite noisy about being hungry, but as they get weaker through lack of milk they will become thin and tucked up in appearance with a staring, rough-looking coat. Prompt antibiotic treatment is required or the sow will lose the use of the infected teats. Feed the piglets a replacement milk and hope the sow's milk returns after antibiotic treatment.

Farrowing Fever – Agalactia

No milk or reduced milk after farrowing. This can have a number of causes. It is most likely to appear after a premature farrowing, especially if the piglets fail to suck or if the sow/gilt is dehydrated or underweight. It can go hand in hand with metritis and is caused by a bacterial infection, so antibiotics are required. It can also be due to the first-time gilt being scared by the piglets or the piglets having torn her teats and they are sore or have been bitten off, and so the let-down response doesn't occur.

The sow/gilt may appear 'normal' and it is the condition and behaviour of the litter that draws attention to the sow's condition. The piglets understandably will be slowly dying and becoming less vigorous, even convulsing. Treatment depends upon the cause, but antibiotics to clear up an infection and an injection of oxytocin to stimulate milk production are a good call, along with copious amounts of fresh water positioned nearby.

NOTE: Metritis, Mastitis and Agalactia are collectively known as MMA as a syndrome and prompt administration of antibiotics is always required.

Torn Nipples/Teats

Piglets are born with needle-sharp teeth, and although it doesn't happen often in outdoor-reared litters, you can get the odd litter that really tears into the sow's teats. This can, understandably, make her reluctant to feed. It can happen at any time but is most likely in the early days when they are scrabbling for teats. Your response will depend on how bad the wounds are and the number of teats affected. If it is one or two teats early on, I would definitely treat the teats as a wound with antiseptic sprays, ensuring they are not toxic to the piglets, and put the sow on a course of antibiotics if they don't start to heal.

If no further damage has occurred, then I may leave it at that. If damage is continuing and/or the sow is still reluctant to feed the piglets, I would on that occasion remove the sow from the pen temporarily and cut the tips off the piglets' needle teeth and then return to the mother.

If the damage starts to occur as the piglets got older, I would be looking at my husbandry:

- Do they have access to water or are they so thirsty they are all but attacking Mum? Am I feeding them enough creep/growers feed so they don't need so much milk to stave off hunger?
- Some breeders do not feed creep feed and just let them pick at Mum's food; if they use sow rolls – the biggest pellet – then small piglets may not bother and so go hungry.

- Do they have enough 'entertainment' to keep them active? Throw in some logs or sections of hay for them to investigate and distract them from chomping on Mum continually.
- Have I remembered to provide water from two weeks of age and they are actually just thirsty? If the piglets are close to six weeks old *and* eating creep/growers feed and drinking water, then consider an early weaning; under that age I would do my best to stop the behaviour with better husbandry techniques, taking the sharp points off the needle teeth as a last resort if I had to.

Prolapse

This involves the complete eversion of both horns of the uterus, which turn completely inside out. It usually takes place within two to four hours of the completion of farrowing but sometimes up to 24 hours afterwards. Prolonged straining during labour can cause a small part of the uterus to be propelled outwards by the contractions before or during farrowing, causing litters to be lost. Unless you have suitable sedatives and powerful painkillers, the sow will be in horrendous pain if you try and extract the piglets, so seek experienced help. Uterine prolapses are reasonably uncommon but usually occur in the older, productive sows with large litters or where large piglets have been born, causing the supporting structures of the uterus to become weak, damaged or flaccid. Treatment involves replacing the womb inside the sow but this is often impossible or the sow dies from internal haemorrhaging. In most cases on welfare grounds the sow should be destroyed.

Anaemia

If you intend to breed pigs, then you need to be aware of anaemia, as it will cause sudden death if you are not careful. Piglets are not born with a sufficient supply of iron, and the sow's milk alone does not provide enough for the growing piglets. Pigs reared outdoors may have access to enough iron in the soil to remain healthy but there is growing evidence that an injection or two of iron in outdoor-born piglets makes for healthier piglets.

The piglet's rapid growth increases the demand for iron and in the first few weeks the ingestion of soil is likely to be by accident rather than design and the soil must be in a chewable form and not baked hard by the sun. Signs of iron deficiency typically start around three to five weeks old and piglets show signs of increasing lethargy, mouth ulcers and pale-coloured faeces; if not treated, they can die suddenly. An iron injection, e.g. Gleptosil™, sweet iron feed sprinkles, or iron paste applied

> **TIP**
>
> If piglets are born indoors or the ground is very hard, provide a freshly dug turf every day to try and prevent iron deficiency. This also acts as 'entertainment' and can help prevent damage to the sow's teats and the piglets fighting. You will still have to remain alert to the possibility of anaemia and have a stock of injectable iron in your medicine box.

to the snout, can be given in the first few days after birth and again, if wished, in the second week as prevention, or as a treatment if you see the symptoms in the first few weeks.

Congenital Defects

Splayed Legs

Piglets are sometimes born with splayed legs and this can affect from one to all four legs. Unless some intervention is applied, then the piglet is unlikely to survive. If all four legs or the two front legs are affected, the prognosis isn't good and it is kinder to have the piglet dispatched. If it is the hind legs only, then the piglet may stand a chance.

You can buy little hobbles secured with Velcro that keep the legs together or you can use electrical tape to bind around their bottoms and legs, incorporating the hips, to keep them together.

If the piglet can get out of Mum's way and feed, then the prognosis is better. What will also help is not having so much straw in the ark that the piglet finds it difficult to get about, or so little that it is slipping all the time. We have had one piglet born like this, applied the hobbles and all was well; he fed with the others and was thriving. We took the hobbles off and the legs were still splaying, so we put them back on. However, over time he wasn't as quick to get the milk as the others and by the time he made it to the teats, he had three suckles and Mum got up. He wasn't thriving at all and so the decision was made to dispatch him.

It can be an inherited condition as well as spontaneous. If the same sow, boar or sow/boar combination continually produces splay-legged piglets, then it's sausages for the guilty party.

Congenital Tremors

It looks like shivering but doesn't stop and is more violent. It is more evident when the piglet is trying to walk or stand and often goes away when it is asleep. It can improve as the piglet ages, but if it interferes with suckling or getting to the teat, then the outcome is not good. It can occur sporadically in litters and there isn't a lot that can be done, but if a sow repeatedly has piglets with tremors, with different boars, then culling her would be a good call. We have had it once in a sow's litter but luckily it wasn't very bad and all the piglets survived. We didn't allow any of them to become breeding stock.

Epitheliogenesis Imperfecta

Epitheliogenesis imperfecta is a failure of the skin to grow and completely cover the body. It most often occurs in patches on the barrel of the pig and on the limbs and is present at the time of birth. If it is of a small size of less than 1.5 cm (0.6 in), then it may be treated as a wound and the skin will slowly heal. Larger diameters may scar but the piglet will most likely make it to pork weight. Where the skin has formed but not joined up, suturing may be a possibility. If it is a very large area that is affected, then the kindest option is to cull the piglet (Figure 9.12).

FIGURE 9.12 Epitheliogenesis Imperfecta in newborn piglet.

This condition seems to be cropping up in my piggy social media feeds more frequently in recent times and I wonder if the condition is on the increase due to inbreeding, as it is thought to be a recessive gene inheritance, or just the people I know are happy to share photographs when it occurs in their herds.

Other Congenital Defects

There are a few other defects to be aware of. Some are spontaneous within a litter, and others are hereditary, due to an infection, nutritional deficiency or a toxin. Spontaneous examples include cleft palate, Siamese twins and heart defects and are relatively rare. Inherited examples may be caused by dominant or recessive genes and include umbilical and inguinal hernias, hermaphrodites, inverted nipples, cryptorchids (retained testicles), atresia ani (blind anus, Figure 9.13) cleft palate, pityriasis rosea (a skin condition of growers) (Figure 9.14), splay leg and porcine stress syndrome (halothane gene positive). Not all the conditions make a difference to the pig destined for the table.

FIGURE 9.13 Blind anus in a boar piglet. Without an operation, he will die, so culling is the kindest option. Gilt piglets may survive to pork weight, as the faeces may pass out through their vaginas.

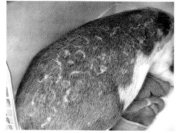

FIGURE 9.14 The classic rings of pityriasis rosea.

We have had one case of pityriasis rosea, also known as false ringworm; it looked vile for a while, with rings of scabby, scurfy skin (Figure 9.14) and she was put on two courses of antibiotics and was fondly known as 'scabbers'. By 24 weeks old you wouldn't have known that she had ever had a problem.

If a sow constantly produces piglets with inherited conditions, using different unrelated boars, then it is likely to be caused by a dominant gene, and your reaction to it depends on the condition she passes on. Personally I would cull repeat offender sows, but you may decide otherwise; of course if because of it you are getting lots of vet bills, and the associated time off work this causes, then a decision to cull needs to be made.

Other Notable Conditions

Piglet Scour

In addition to the causes of scour mentioned in Chapter Five, piglets can sometimes get a mild scour at aged three weeks. It is thought to be the reaction to the sow having a 'quiet season' and the associated hormones that go with it. I have never noticed this with our sows but I know people who have.

Joint Ill

Joint ill is an infection, most common in young piglets around two to three weeks old, that enters the bloodstream through an open wound – even one you can't see or the unhealed navel – and the infection materialises around the joints as large balloon-like swellings and the piglets can become very shaky when moving.

Do not lance the infection, as you will just create another wound. The piglets may or may not be lame, depending which joints are infected and how early you spot it. Treat with broad-spectrum antibiotics for the recommended length of time on the instructions and the infection should go away with no lasting damage if you treat early enough. If you do not have antibiotics, then you will need to call the veterinary surgeon to administer them. We have only ever had this with one sow: in every litter she had with us a couple of piglets succumbed, requiring antibiotics. We took the decision to cull her, but it is a treatable condition in the piglets.

Greasy Pig Disease

Greasy Pig Disease, also called exudative dermatitis, is caused by the bacterium *Staphylococcus hyicus*, which invades broken skin causing an infection (Figure 9.15). The bacteria produce toxins that may damage the liver and kidneys in even a moderate infection, and not all survive, especially nursing piglets. Characteristic dark brown skin lesions where the skin has been damaged first form near the initial site of infection, then spread around the body. The skin becomes wrinkled with large areas of flaking over the body, with a greasy feel to it. In weaned pigs disease may appear two to three days after weaning, with a slight browning of the skin that progresses to a dark greasy texture, and in severe cases the skin turns black and the pig usually dies. If the sow has passed on some immunity, then the infection may stay localised, with small lesions that do not spread. In nursing piglets

FIGURE 9.15 The classic flaking of the skin in Greasy Pig Disease.

the disease is usually confined to individual animals, but it can be a major problem in farrowing gilts' progeny and weaned pigs. Antibiotic treatment early in the infection will help, both systemic and topical, and washing the affected pig with anti-microbial washes, e.g. Hibiscrub™. If it is a major problem in weaned pigs, then further veterinary investigation and intervention is required.

Bowel Oedema (Salt Toxicity) in Recently Weaned Piglets

Bowel oedema is difficult to avoid, but in reality it is not that common – we have had two cases in the last ten years and one potential case. What usually happens with toxigenic bacterial poisoning is that you find a good pig dead one morning, most often the largest in the litter, typically with the eyelids and face appearing swollen.

Bowel oedema is caused by a pig ingesting toxigenic *E. coli* bacteria from the soil while it's rooting about. In a susceptible pig the bacteria proliferates in the gut, which produces a toxin. This toxin blocks the absorption of water in the large intestine, effectively causing dehydration no matter how much the pig drinks. This dehydration causes salt levels in the brain to become dangerously high, eventually leading to paralysis and death. Antibiotics in this situation, even if you had time to administer them, would escalate the onset of death, as when the bacteria die, they would all release any toxins at once.

The condition is very rare in the suckling pig as the immunoglobulins in the sow's milk are highly protective. It is most common in the month after weaning and the withdrawal of the protective milk, with days 10–14 post-weaning the most likely time it will occur. It is a primary reason why responsible breeders do not pluck a piglet off its mother and sell it on to a new home immediately. The prognosis is not good and the condition will quickly worsen, so humane dispatch is required. I did try on all three occasions we have experienced it to feed whole milk via a syringe in the hope of blocking the toxin; one lived, two died, but I don't know if the one that lived had the condition and I saved it . . . I suspect it didn't.

Can You Breed Pigs and Work Full-time?

THE FINAL JOURNEY

Around 10 million pigs are slaughtered annually in the UK and 110 million in the USA, with the welfare of each one, quite rightly, protected by law. Large commercial producers and small-scale pig keepers all have to follow the same basic legislation. In the UK, legislation is covered by two laws: The Welfare of Animals (Slaughter or Killing) Regulations 1995 (WASK) and The Welfare of Animals at the Time of Killing (WATOK). In the USA, the Humane Methods of Slaughter Act (HMSA) governs the welfare of pigs at the time of slaughter.

WHO MONITORS ABATTOIRS?

The Food Standards Agency (FSA) is responsible for the licensing of abattoirs within the UK and the USDA Food Safety and Inspection Service (FSIS) in the USA. They employ state-approved or Official Veterinarians (OVs) and Meat Health Inspectors (MHIs). OVs are responsible for all welfare issues before and during the pig's time at the abattoir. This means anyone arriving with a pig showing evidence of poor welfare, on-farm and/or during transport, may be reported to the appropriate enforcement body. MHIs are responsible for ensuring that appropriate procedures for the handling/chilling/identification of meat are upheld and that all carcasses are fit for human consumption. USDA FSIS inspectors and MHIs are required to stamp an official seal of approval on meat they have passed as fit for human consumption.

By law, pigs must be spared any avoidable excitement, pain or suffering during movement, time in the lairage, restraint (if practised), plus during the stunning and subsequent killing. To obtain the required licences, the staff at the abattoir must have proved themselves competent and have the skills and knowledge to perform all tasks humanely.

FINDING AN ABATTOIR OR PLANT

Slaughterhouses, plants or abattoirs for small-scale producers are getting harder to find but they are out there. Some are better than others and some have a better reputation than others. The FSA website lists all the abattoirs in the UK and the FSIS website lists those in the USA; both list the type of species they are licensed to slaughter.

In the UK, all licensed abattoirs are inspected and the meat can be sold anywhere in the country, subject to complying with FSA meat hygiene regulations. In the USA, where you can sell your meat may be determined by the type of abattoir you have used. To sell the meat interstate, by direct sale or post, then you must use a federal-inspected abattoir, a Talmadge-Aiken (TA) abattoir or federal-inspected mobile slaughter unit (MSU). If you only wish to sell within state, then you can use any federally inspected abattoir, a state-inspected MSU or a state-approved abattoir. There is some derogation from these rules if you are a member of a co-operative interstate programme.

Finding an abattoir with a good reputation is important but not always easy. Ask other farmers or try social media forums. Abattoirs in both the UK and USA are licensed for specific breeds of animal, so check which they take, confirm that they take small numbers of animals, and ask which day of the week they take specific breeds. Don't forget your nearest one might not be the best one, although in the USA it is not uncommon to have to cross state lines just to find one at all.

There are certain times when you will need to book ahead. Christmas and Easter are obvious, but additionally in the USA, June to August are also busy periods due to state fairs and carcass competitions and also Thanksgiving.

Cull Abattoirs

There are specialist abattoirs in the UK and USA which will buy any type of pig, including fully grown sows and boars, from you once they have finished their working life. These pigs are known as cull sows and boars and it doesn't matter how big they are; in fact, you get more money the bigger they are, as it is all paid by weight. The meat is graded at these abattoirs using a variety of optical probes and fat monitors and you get paid the current cull price per kg (lb) for the grade achieved.

I was told by my abattoir that you can legally take pregnant sows to the abattoir up to three weeks before their due date, but the abattoirs understandably don't like it. I can only find UK WATO travel requirements on the final 10 % of gestation and first week after farrowing in the WATOK regulations. Whether it is 11.5 days or three weeks, it doesn't sit too well with me either, and if I was getting the carcass back, I doubt I would enjoy the meat, thinking about the piglets.

WHEN ARE MY PIGS READY FOR SLAUGHTER?

You can of course send your fattening pig to the abattoir pretty much whenever you want. In slower-maturing traditional and heritage breeds, the pigs are around six months of age for pork and eight months of age for bacon, although some breeds, like the Berkshire and the Middle White, are often sent at five months for pork. Traditionally, bacon-weight pigs are larger to provide a better-sized 'eye' on the back bacon rasher and more ham/gammon joints, but it's your pig so it's up to you. You can eat the meat as pork at bacon weight: you just get a bit more of it and I doubt the meat becomes tougher in a month or two.

Most abattoirs don't like pigs that are too big and may charge a higher price for larger pigs as it takes longer to scrape the hair off, so if you send in very large pigs, they may insist on skinning them. Abattoirs in the USA usually deskin the pigs, but a growing number now offer dehairing as an option.

We send our British Saddlebacks to the abattoir at around six to seven months old for pork joints, chops and sausages. At this age they weigh between 60 and 65 kg (130 to 145 lb) dead weight – the weight of the carcass after the innards have been removed – which is approximately 72 % of the actual live weight of 83 to 90 kg (180 to 200 lb). This is often referred to as the 'kill out' percentage (KO %). Our pigs destined for bacon and gammons are aged from seven to nine months at approximately 65 to 75kg (145 to 165 lb) dead weight. This may be earlier or later than some other small-scale keepers, but the size we get back suits us.

While a KO % of 72 to 75 % is pretty much average for the traditional and larger heritage breeds, some can realise a KO % of up to 90 %, as reported in the Middle White and Berkshire breeds, especially if you send them at slightly earlier ages, as boasted on their promotional material. That means quite a lot more bang for your buck. Our Middle Whites do have a higher KO % than our British Saddlebacks, although not as high as 90 %, but then we are not 'proper' fattening pig producers and have not finely tuned our feeding and rearing practices to suit the butcher.

ESTIMATING THE WEIGHT OF YOUR PIG

There are various ways to estimate the weight of a pig. A weigh crate will give you the pig's actual live weight, and knowing your potential KO %, you can estimate the dead weight you will get back. A weigh tape can be purchased, and most give you a dead-weight reading. Measure the pig just behind its front legs and read off the weight.

Alternatively, measure the circumference of your pig just behind its front legs in centimetres/inches with an ordinary measuring tape and then read the approximate weight off the chart below (Table 10.1).

TABLE 10.1 The approximate live and dead weight of a pig from measuring the pig's circumference behind the front legs.

CIRCUMFERENCE OF PIG		LIVE WEIGHT OF PIG		DEAD WEIGHT OF PIG	
CM	INCHES	KG	LBS	KG	LBS
88	34.5	74	163	53	117
90	35.5	78	172	56	123
92	36	81	179	58	128
94	37	86	190	62	137
96	37.5	92	203	66	146
98	38.5	97	214	70	154
100	39	101	223	73	161
102	40	104	230	75	165
104	41	107	236	77	170
106	42	110	243	79	174
108	42.5	113	249	81	179
110	43.5	117	248	84	185

Remember that if you have a commercial breed or a cross breed, then they may reach pork or bacon weight quicker. Also some traditional breeds mature faster than others; the above ages I have quoted are what we use for our British Saddlebacks as a guideline. You may also need to bear in mind the conformation of your pig: if it has massive hams compared to the shoulders, then your estimated dead weight might be less than the weight you actually get back, if you used a tape.

BEFORE YOU TAKE YOUR PIGS TO THE ABATTOIR

So, your pigs are booked in to the abattoir and your time slot allocated, your butcher is ready and waiting and all you need to do is get your pigs into the trailer and be on your way. In the UK, complete your movement licence and print off your haulier sheet to take with you or give to your haulier if you are using a third party. In the USA, if the abattoir is interstate, then you will need a movement form signed by your state-approved veterinary surgeon.

You will need to have a clean, disinfected livestock trailer for pigs of this size and you may be turned away if it is not suitable. So buy, beg or borrow if you have to. If you have your own trailer and the pigs are not scrunched in, then you can load them the night before, if you provide them with bedding and water; this way you will not be going over your permitted eight hours in a trailer, as it can be called a pen. The advantage of this is that you can use food to persuade them to go up the ramp; if you load on the morning of slaughter, then you cannot use food because the abattoir does not want too much undigested feed in their system. This is not because of extra to remove, but the reduction in the volume of gut contents means fewer bacteria, which reduces the risk of contamination of the carcass during dressing (splitting the carcass and spray washing to remove blood etc.).

TIP

If it is a long distance you need to walk the pigs, it has been shown that they will happily follow curved and straight paths, so avoid sharp angles in your route and they are less likely to run back on you at full pelt.

You will also need to ear tag or slap-mark the pigs for identification. I think the day or two before slaughter is most acceptable and doesn't add to the pigs' stress on the actual morning. The pigs are definitely more relaxed about the whole affair and often remain asleep until they get there.

If your pigs are very muddy, it is a good idea to bring them in to dry off a couple of days before you go, or at least move them to a less muddy pen in the hope they will become a bit cleaner. They do not need to be immaculate, just not caked in mud. This is mainly for the pigs' benefit, as if they have a muddy head, the electric stunning may not be as efficient.

If you decide to load on the morning of slaughter, then 'Sod's Law' dictates that your pigs will not load quietly and quickly, so allow yourself plenty of time. If they do load quickly, there is no harm in allowing them to settle in the trailer for a short while before you set off. You may think it is a good idea to have the trailer in the pig's vicinity for a few days before loading, and even feed them in it to get them used to loading; you can, but they will eat the electrics and the tyres. As described in Chapter Six, practise walking them out of their pen and to the trailer point, until they are used to it.

Keeping Stress Levels to a Minimum

It is not a good idea to pump the pigs full of adrenaline and excite them by chasing them about, or any other scary practice. Stress through fear or excitement causes the pigs to overheat, which can have a major effect on the quality of the meat. Stress prior to slaughter can be caused by chasing them about prior to loading, fighting, rough handling and electric-prod use. Stress is a major contributor to the release of stored glycogen (sugar) into the bloodstream. The 15 minutes prior to slaughter is thought to be the most important but it can take a stressed pig a few hours to calm down. After slaughter this sudden release of stored glycogen causes a partial breakdown in the muscle fibres and low acidic pH, resulting in a pale abnormal colour and dry consistency to the meat, rendering it tasteless with reduced water-holding capacity when cooked, and is known as Pale Soft Exudate (PSE), and at a lesser level called Red Soft Exudate (RSE), where only the texture is affected.

Additional causes of increased susceptibility to PSE and RSE include the breed of pig known to carry Porcine Stress Syndrome (PSS) genes (see Chapter Eight), slow carcass chilling at the abattoir and warm weather. In the USA it is considered so important, that to market your meat through some Premium Quality Meat schemes, the pigs have to be tested for the gene prior to slaughter to minimise the chance of it occurring. Due to the high acidity, the meat doesn't spoil as quickly from microbial contamination, and so if that pack of chops you forgot about at the back of the fridge is still edible after a few weeks, it might be the reason why!

Stress can also induce Dark, Firm, Dry (DFD) meat, although this is typically associated with ruminant species and rarer in the pig. However, all muscles of the pork can be susceptible, especially if the pig is stress gene positive. DFD muscle is dark in colour and less appealing to consumers, who often interpret it as an indication of meat from older animals, or lacking in freshness. DFD meat is more susceptible to microbial spoilage because of its exceptional water-holding ability and its higher pH (less acidity), which favours bacteria growth.

WHAT YOU NEED TO DO ONCE AT THE ABATTOIR

Once at the abattoir, wait until you have permission before unloading. Your pigs may be reluctant to unload, so make sure you have straw on the ramp and climb inside the trailer to gently push them out. You'll need to be efficient because otherwise the lairage man will do it and he will be quite firm – he can't afford to stand there watching you gently tickle them awake and coax them out.

Once unloaded, you can now move your trailer away from the doors and park where instructed before sorting the paperwork out in the office. You'll be asked to confirm what is happening to your carcass if the abattoir isn't performing the butchery. Most abattoirs allow you to pick up your carcass in the back of your car if it's solely for your own consumption. If you're selling the meat to a third party, then you'll require refrigerated transport to the point of butchery or customer, which some abattoirs provide for a fee. Before leaving the abattoir you'll also have to clean and disinfect your trailer or sign an 'undertaking to cleanse and disinfect' form, provided by the abattoir, and clean at home within 24 hours or before next use.

WHAT HAPPENS ONCE YOU HAVE UNLOADED YOUR PIGS?

This is the only part of the pigs' lives that you have no control over and so to me is the part that worries me the most. Selection of the best abattoir from your pigs' point of view is not easy. Most are reluctant to allow you to watch, because of the behaviour of some animal rights activists and biased reporting when people do gain access. I'm not saying that awful practices don't happen; we have all seen horrific videos of unintelligent abattoir workers bullying pigs in a barbaric way and this quite rightly must be eradicated. The problem comes with people taking pictures of the process; pictures can be subjective, and what looks horrific in a photo may not have been a welfare problem to the pig at all, e.g. it was unconscious at the time. This is the reason it was difficult to get photographs of the process for this book.

What the activists have effectively done is to prevent openness between abattoirs and smallholders to allay their fears about their pigs' last moments, which is truly a shame. The abattoir we use is a small non-commercial enterprise and the reception is another room in the same building, so believe me, if a pig was squealing in pain or distress I would be able to hear it; not once have I heard a sound and I have to take comfort from that.

Once unloaded, the pigs are guided into an empty pen inside the lairage, eliminating fights with unfamiliar pigs. The OVs then assess their health to ensure they're healthy enough to enter the food chain (Figure 10.1). The pigs are held for as short a time as possible and are probably already 'gone' by the time you're on your way home. In larger abattoirs they can be held for 24-plus hours with access to water to flush out their digestive system.

The pigs are guided in their groups through to the 'processing room' where they're quickly stunned with an electric shock, rendering them unconscious by inducing epileptic-type brain activity, similar to *grand mal* epileptic fits in humans, and unable to feel pain. The pig doesn't have time to squeal and the other pigs will just assume it's lying down.

There are two basic types of electrical stunning equipment used for pigs: head-only and head-to-body stunning. In head-only stunning, hand-held scissor-like or rigid tongs are placed on both sides of the animal's head so that the electrodes span the brain (Figure 10.2).

FIGURE 10.1 *Pigs being inspected by an official veterinary surgeon at the abattoir.*

Photo courtesy of Liz Shankland.

FIGURE 10.2 Head-only electrical stunning.

Abattoir personnel are trained to know the signs of an effective stunning, which include the pig collapsing with no rhythmic breathing and becoming rigid with its head slightly raised, its forelegs extended and its hind legs tucked into the body. This is followed by a gradual relaxation of the muscles accompanied by often violent paddling motions and the eyeballs rotating downward. The pig may also defecate and/or urinate.

The head-to-body process involves holding the animal in a 'restrainer', which looks similar to a pig stall. Electrodes are placed over the head and back, passing current through the brain and heart, simultaneously stunning the animal and causing a cardiac arrest and instant death. Abattoirs can also use CO_2 gas to render them unconscious, the advantage of this being that large numbers of pigs can be done at once and without the additional stress of human contact.

The pigs are then shackled upside down by their hind quarters and 'stuck', which involves the slicing of the carotid artery and the jugular vein; this allows blood to freely flow into a drain (or container if you wish to make black pudding). The pig bleeds out for a minimum of five minutes, ensuring death and allowing the muscles to relax before further processing.

The pigs are then dehaired, usually via a scalding process, where the carcass is put through a water bath of 60–68°C (140–155°F), opening the pores in the skin and loosening the hair. The pig is then scraped with a blade until all the hair is removed and stubborn hair removed with a flame. In the USA it is common practice to remove the skin, although some abattoirs will scald and scrape. The carcass is then washed with water to remove 'the bits' and then eviscerated (the removal of the internal organs). The majority of the internal organs are disposed of, but you'll get back the edible heart, kidney and liver if they pass the meat health inspection (Table 10.2).

The MHIs inspect the offal for any signs of disease, including the liver for 'milk spot' – the scarring caused by the *Ascarid* worm – and for things like peritonitis. Certain joints will also be examined and removed if required. Any abscesses in the meat, caused for example by bad injection techniques or injury, will also be excised or that particular part of the pig condemned. Once passed as fit for human consumption, the carcasses are stamped with an official stamp (Figure 10.3)

Assuming all is well, the carcass is then split into two halves, sometimes with the complete head attached to one half, and washed in a mild salt solution to remove blood, bone and organ tissue, before finally being chilled down to 4°C, ready for collection/ butchering. If you wish to use your pig for a hog roast, don't forget to tell the abattoir, so they don't split it in half!

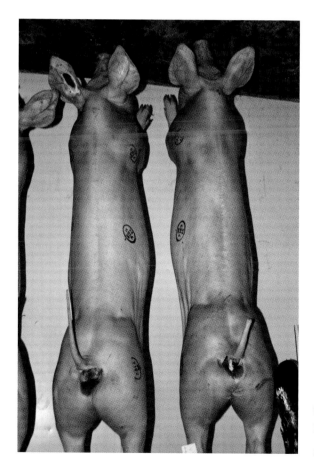

FIGURE 10.3 The Meat Health Inspector's official stamp.

Photo courtesy of Liz Shankland.

TABLE 10.2 *The most common parts of the pig that are removed from or retained on the carcass – some parts are returned separately to the carcass, others you may have to request back.*

BODY PART	REMOVED FROM/RETAINED ON CARCASS
Hair	Removed
Liver/heart/lungs	
Guts	
Caul and gut fat	
Genito-urinary organs (minus kidneys)	
Scrotal sac (boars)/udder (sows)	
Head	Retained
Ears, ear root and eyes	
Neck (minimal trimming to neaten is allowed)	
Tongue	Either
Flare fat, kidneys and diaphragm	(depends upon abattoir)
Cleys on trotters	
Tail	

CARCASS GRADING

Some of the larger abattoirs have trained assessors, e.g. in the UK, Meat and Livestock Commission (MLC)-trained assessors, who can evaluate the carcass using various optical and grading probes for lean muscle quality, amount of back fat, amount of internal fat marbling, etc. The measurements obtained provide an overall carcass-quality score. If you wish to know what the paperwork means, the MLC have an easy to understand document on their website.

Currently in the USA, pork is not graded with USDA quality grades, their reason being that pork is generally produced from young animals that have been bred and fed to produce more uniformly tender meat. Premium Quality Meat schemes may, however, use a grading system. Grading is an optional service, unlike the Meat Health Inspections, which are mandatory in both the UK and USA.

CARCASS COLLECTION

Some abattoirs allow you to pick up your pig yourself in the back of your car and others insist on delivering your pig in refrigerated transport to their destination for a fee. If you are butchering your pig solely for your own consumption, then the back of your car is fine, especially if it is not very far away. If it is the height of summer and you have a fair distance to drive, the use of large coolboxes or a heat-resistant wrapping material would assist in keeping the carcass or butchered joints from spoiling. If you will be selling the meat to a third party, then you will have to have the pig delivered to yourself or your butcher by refrigerated transport. If you are butchering yourself and will be selling the meat on, you will need to have where you are butchering your pig approved by Environmental Health, and there are some other hoops you need to jump through as well.

SOURCING BUTCHERS

Often abattoirs have a cutting (butchery) service attached to them and will provide from a basic level (slapdash sometimes) to high-end butchery, depending on where you go, with the advantage of not having to transport your carcass to the butcher. Some, however, do not have a curing service, so you will have to source a butcher who does this or try doing it yourself. If you have chosen a butcher to do the cutting, you will also need to check who will transport your pigs to the butcher and how much it will cost per pig.

Word of mouth is a good way to find a reasonably priced butcher. Ring around local butchers and ask if any provide the service you're looking for. Some butchers have a relationship with abattoirs and arranging delivery will be sorted instantly. Ask your butcher for a convenient time to discuss the cuts you want, if you want sausages ask what flavours they do, and if you can have smoked or cured meats. This may mean two or three trips to the butcher to pick up joints, then sausages, then cured meats. Pork is rarely hung for more than a couple of days before cutting, and when it is, it is for the convenience of the butcher rather than a need for the meat, but sausages usually hang for 24 hours to dry before being packed. Your butcher will let you know when to collect your meat, which is useful if you in turn have customers waiting.

NOTE: Be warned that some butchers can charge like an angry bull for the same level of service and expertise of another – so talk money before you have the pigs delivered. I also like to try some of the products the butcher has for sale before even asking.

PRIVATE 'ON-FARM' KILLING REGULATIONS

If you find this all a bit gamey and would rather slaughter them at home, then you are allowed, under strict criteria, to do so. Pigs slaughtered on the farm benefit from the same WASK and WATOK regulations as abattoirs and it is still an offence to cause or permit any avoidable pain, excitement, distress or suffering to the pig.

There are books detailing the best methods to shoot pigs, but unless you are both confident and competent, you could easily fall foul of the regulations, and the very stress you are trying to avoid for the pig actually made worse.

You will also need to have suitable buildings to be able to winch the pig up quickly in order to drain the blood and you will have to know how you will dispose of the non-edible parts of the carcass. Even the fashionable 'nose to tail' eating must leave one or two parts you or your dogs would rather not eat, e.g. intestinal contents.

Your local fell-man with a licensed incinerator will take them for a fee. In the UK, if you employ the services of a licensed slaughterman to both kill and process your pigs on the farm, and then you consume the resulting meat, you run the risk of contravening EU food hygiene regulations. This is because, by law, a dressed carcass cannot be supplied for consumption unless it has been subject to inspection and health marking by the Meat Hygiene Service in licensed premises. If you employ the slaughterman to only kill your pigs and he/she will not be dressing the carcass (removing the innards, etc.), then this may be legal, but the rules are rather unclear.

In the USA, in order to support remote rural economies and assist the local produce suppliers from having to travel often hundreds of miles, the USDA FSIS has granted the use of federal-licensed mobile slaughter units (MSU), which can travel from site to site by appointment. The meat from these units may be sold locally or interstate as a Federal Inspector travels with the unit. The number of pigs that can be processed is often limited by the size of the chiller in the unit, but between 10 and 25 pigs is a common number that can be accommodated. All pigs are still protected by the Humane Methods of Slaughter Act, covering their welfare, and their meat health inspected.

Another option in the USA is Custom-Exempt meat, which is killed on farm but not routinely federal or state inspected. This meat can, legally, only be consumed by the owner, his/her immediate family and non-paying guests. However, if a customer buys the pig live, then it can subsequently be killed on-farm and be consumed by their immediate family.

YOUR FINAL GOODBYE

For most small-scale fattening-pig keepers, the abattoir is the final destination. You may find that you feel sad when the pigs leave the trailer and go off into the lairage, and I think that is the way it should be. If you feel something, then it means you have genuinely cared for them when they were alive. Of course if you're hanging off their legs, sobbing and crying as they go in, then perhaps fattening-pig rearing isn't for you.

HOW MUCH IT ALL COSTS?

The costs involved are, of course, linked directly to the size and complexity of your piggy enterprise. I deliberated about including prices within the book as they will of course rapidly go out of date, but it gives you a starting point, and a few phone calls or Internet searches will give you the current price. The costs of rearing a couple of weaners to pork weight are significantly less than keeping a herd of breeding sows, but I will try to cover most angles and then you can multiply up to the number and ages of pigs you wish to own and for how long you will have them.

SETTING UP YOUR SYSTEM

The cost of setting up your system can be divided by the number of times you will use it over the years and cannot be attributed all to your first pigs. I will be describing the equipment we use to successfully keep our pigs contained. We know people who had much cheaper setting-up costs using no stock fence, just electric wire or vice versa, and it's probably no coincidence that these are the people who constantly moan about their pigs getting out all the time, running riot and have brother–sister mating 'unexpectedly'. If a fertile young female can reach the boar . . . she will! Yes, that is correct, the sow will be the one who goes searching for the boar, although he is always happy to oblige her less than ladylike demands.

You can of course use existing buildings, walled gardens and fencing to keep costs down. Secondhand pig arks, water and feed troughs are available, but cost in buying a powerful disinfectant to clean them with. Occasionally you will see secondhand energisers for sale, although never when you need to buy them, but we have found that they have

> **TIP**
>
> As a *Commuter Pig Keeper* you just cannot afford not to have belt-and-braces fencing and need to know where your pigs are at all times.

TABLE 11.1 A guide to set-up costs in 2016.

ITEM DESCRIPTION	£ 2016	$ 2016
Housing: 8 x 6 ark plus floor	£300–£500 £450–£600	$750–$1200
Hose pipe	£50	$40
Water troughs (each)	£30–£150	$60+
Feed troughs/bowls (each)	£20	$15
Fencing: Wooden posts (each) Wooden strainer posts (each) Stock fence (50 m/165 ft) Stock fence (100 m/330 ft)	£1.50–£3.00 £10 £40 £70	$3.00 $12 $60+ $80+
Post rammer/wire strainers	£40/£15	$50/$25
Electric insulators Electric handles (each)	£20 per 50 £5	$12 per 25 $5
Electric wire or tape (100 m)	£25–£35	$25–$40
Plastic push-in temporary electric posts (each)	£3	$5
Energiser	£100–£300	$150–$400
Leisure battery or two	£65	$100

a limited life span of approximately three to five years and so they may be a false economy if they are old.

DIRECT COSTS PER PIG TO PORK WEIGHT

The costs after setting up your system are the direct costs per pig. Some of the costs are unavoidable but some can be negated, e.g. if you have a trailer or access to one and don't have to hire, and whether you can butcher the meat yourself.

The cost of buying an eight-week-old fattening/feeder weaner ranges from around £45 to £65 ($50 to $200 in the USA), depending upon breed chosen and where you live. You will see piglets advertised much cheaper and you may be getting a bargain, but you may also be buying a heap of trouble. That is a judgment call on your part.

You will need approximately five bales of straw (more in the winter as they tread mud into the ark all the time). This cost can be spread over a few pigs living together.

You will need approximately 275 kg (600 lb) of pig feed per pig to pork weight and around 400 kg (880 lb) per pig to reach bacon weight. The price of pig feed fluctuates according to the world's demand for grain, and crop yields, but the trend is for them to rise rather than fall. However, this could be an area where you can save if you have more than a few pigs and can buy and use larger quantities of feed before sell-by dates. Substantial savings can be made from buying in bulk, usually a tonne at a time, more if you can store three or more tonnes of feed loose in a bulk container, and substantial savings if you can grow your own and have the time, knowledge and storage facilities to process it. Ask your feed merchant for bulk discount sales.

A typical 25 kg (50 lb) bag of 13 to 16 % protein sow and weaner nut or roll costs (2016) between £6 and £11 ($15), but this will vary by region, outlet and quantity purchased. Starter/creep feed and grower pellets, between 18 % and 23 % protein, can cost almost double at £15 per 25 kg ($27 per 50 lb) sack, but if you use it, you will be feeding correspondingly less of it and for a short period of time. There are organic and GM-free feeds available in both the UK and USA across the feed and age ranges. They often come in smaller pack sizes of 20 kg (40 lb) and cost more, but if you are selling organically reared pigs, then you have no option. In 2016 for creep/grower expect to pay an average of £11 ($30 to $50), sow and weaner £10 ($25) and finisher £13 ($30).

Then you will have abattoir fees, which are from £25 to £30 ($45 to $100) per pig. This includes the killing of your pig, removal of its insides, removal of the outer hair, veterinary inspection charges before slaughter, Meat Health Inspector charges after slaughter and the splitting of your pig into two manageable halves. The splitting charge is removed if you keep your pig whole, e.g. for a hog roast.

BUTCHERY OR CUTTING FEES

Some abattoirs also have a cutting house (butchery) attached and so you can save the transport costs from the abattoir to the butcher. Sadly our first experience of butchery was at the abattoir's cutting house, and to say it was a bit slapdash is an understatement: the leg joints we had back would never have fitted in an oven and the packaging was a plastic bag with a sticky label that I believe – before it fell off in the freezer – simply said 'leg', and the 20 kg of sausages we had made came in one bag! Not all cutting houses attached to abattoirs are like this, and the one we used many years ago may now be fine, but do your research to see if the one you could use is any good.

We now have our carcasses transported to a professional butcher's and pay that bit extra for vacuum-packaged meat. So unless you will be butchering yourself, you will also need to add in the butchery costs, this is circa £0.65 to £1.00 per kg ($0.55 to $0.80 per lb) to cut into joints, add £2.50 to £3.50 per kg ($2.50/lb) to make sausages, or £1.70 to £2.70 (additional $0.60/lb) to cure joints for gammon or for bacon, and then £2.50 to £3.50 ($3/lb) for slicing the bacon. Check these fees include high-end vacuum packaging, especially in the USA where the butcher often just wraps the cuts of meat in butcher's paper. Vacuum packaging helps prevent the meat from spoiling when frozen and makes it easier to sell the meat on if you wish.

BUTCHERING AT HOME

If you are not selling the meat, you can save all the butchering costs and butcher the pig yourself. Invest in a cleaver, a boning knife, a sharpener and a saw. We use various websites and posters to help us identify which bits are which, and although the results are not as neat as a butcher's, it tastes just as delicious. The website www.lovepork.co.uk is a fantastic source of knowledge and you can even order posters free of charge to help you do the job. YouTube is another valuable source of information and there are many books on the subject, most with detailed diagrams to help you. At the time of writing, if you are in the UK, then Quality Meat Scotland also produces a free butchering DVD.

WHAT ARE THE POPULAR MEAT CUTS?

What joints and cuts should you choose first time? We had no idea what we or our customers wanted, so we asked the butcher (omitting the cutting house fiasco) for some manageable joints of 1.5 kg (3.5 lb), different types of chops and some sausages to see what we thought. For us, joints and sausages are the best sellers, with sausages the outright winners, and at Christmas our smoked and dry-cured gammon joints sell out very quickly. Remember gammons are usually from bacon-weight pigs and so can be up to nine months old when slaughtered. Add in the butchering and then curing times of around four weeks, and you'll see that you need to buy piglets born in January/early February of the same year to get them back as gammon in time for Christmas.

NOTE: Fuel, electric, water rates, veterinary costs, etc. have not been included.

TABLE 11.2 A summary of direct costs.

TO RAISE AN AVERAGE PIG FROM EIGHT WEEKS TO PORK WEIGHT	£ 2016	$ 2016
Cost of weaner	£50	$100
Straw x 5	£15	$25
Feed x 11 sacks	£110	$165
Slaughter fees	£30	$70
Butchering costs (basic) Joints and sausages Vacuum packaging	£60	$100
Total	£265	$460

HOW MUCH HAS THE MEAT COST YOU TO PRODUCE?

Or if you wish to work out price per kilo to produce the meat, in order to calculate what to charge, then if you get 65 kg (140 lbs) dead weight back from your pig, lose 10 % for the head and some of the bones that will be removed, depending upon how you have it butchered, so if it's all sausages, then you will lose much more, which leaves 58 kg (128 lb). That 58 kg (128 lbs) has cost you a minimum of £205 ($360) to produce in total and so means every kg (lb) of meat has cost £3.53 ($2.80). You now have to cost in the butcher's cutting fee (£0.80/kg or $0.68/lb), so that has increased to £4.53/kg ($3.48/lb) to produce.

If you then cost in sausage making, curing, bacon slicing and charcuterie, you can see how quickly the invoice can add up. While sausages and bacon may be best sellers, it's clear they cost an awful lot more to produce than our joints.

You will also need to cost in fuel to pick up feed, take the pigs to the abattoir and transport the carcass from abattoir to the butcher; electricity costs for your battery recharger, lighting, etc.; water if you are metered; your labour time if wished, and a vet bill if you are unlucky.

The curing and smoking of meat by your butcher, and especially slicing of bacon, adds quite a lot to the final invoice but is so worth it. I prefer smoked and Neil prefers dry cure, so try both if you can. To make money from your meat, you have to sell to the end user; butchers may well buy your meat but you will not even recoup your costs. We don't have time to go into meat sales in a big way and so have modest experience in this area. We pitch our meat prices to match the premium ranges of the better supermarket ranges; this way people are more likely to buy the meat, although check out the famous River Cottage or Duchy of Cornwall price lists: you would be amazed at what people will pay for good-quality meat.

You can now see how if you sell the meat from two pigs at a decent but fair price for the quality, you can pay for all the meat from a third pig. For example, if we had 58 kg of meat joints, selling them for an average of £9 kg equals £522. Less our costs of £265 equals £257 profit. A number of smallholders sell enough to pay for all their immediate costs, a little from off the set-up costs and then enjoy their 'free' meat.

The cost per piglet when you breed yourself can be, but not always, substantially lower than if you buy them at eight weeks. However, the costs prepared in Table 11.3 assume no vet bills, modest- to large-sized litters or no other unusual expenditure, such as farm health plans or keeping the sow empty between litters. The table does include the cost of feeding the sow – including the extra for flushing her before mating and feeding piglets for two litters per year; an allowance for the boar's keep (remember for these calculations boars get cheaper the more sows you have); bedding costs; and deworming and vaccination. I have also calculated the different cost for gilts purchased at eight weeks and reared to 12 months before covering, and for sows that have already

TABLE 11.3 *An example of the cost of producing your own piglets.*

NUMBER OF PIGLETS IN LITTER	COST PER PIGLET FIRST LITTER	COST PER PIGLET SUBSEQUENT LITTERS
5	£60	£30
6	£50	£25
7	£44	£23
8	£40	£21
9	£36	£19
10	£33	£18
11	£31	£17
12	£29	£16
Cost of whole litter to produce	£302	£152

had one litter. This is important, depending upon the age you buy them, but I have not included the purchase price of the sow or gilt.

NOTE: All the prices quoted in pounds sterling or US Dollars are the extremely approximate costs in 2016 and vary from region to region and in time; no emailing me, because it cost double what I have quoted, especially if you are reading this book in 2032! The table is designed to give you a ballpark figure for possible expenditure, not an absolute cost.

It also gives you an idea of how someone who is selling weaners at ridiculously low prices must be cutting corners somewhere, otherwise piglet production is just not sustainable.

MARKETING YOUR EXCESS STOCK

It is likely, unless you have a thriving hog roast business, own a butcher's shop or you breed for the local butcher, that you will need to sell some of your home-bred weaners.

MAXIMISING YOUR PROFITS FROM SELLING PIGS

Maximising profit means providing the best-quality product for sale, for the cheapest cost to produce. That doesn't mean cutting corners but rather starting with good, strong, quality breeding stock, keeping them healthy via deworming and vaccination programmes, and maintaining high standards of biosecurity and husbandry. The choice of breed can be one way of ensuring a strong customer base willing to pay a premium. Breeds seem to become 'fashionable' in waves and, at the moment, newcomers to pig keeping in the South of England seem rather partial to the Oxford, Sandy and Black breed. If you are undecided and haven't yet purchased your starter herd, then you could choose a breed not yet in the area, securing yourself pole position in a niche market. Of course you have to source breeding stock from further afield and so, due to diesel costs, you will pay more, but this can be negated by the profits made. If you are going for a more commonly available breed such as the British Saddleback or Gloucestershire Old Spots, you might want to choose one of the more prolific bloodlines of the breed to maximise the number of piglets born and raised.

Alternatively, you could concentrate on the rarer bloodlines and possibly exploit the market when selling young and older breeding stock to those who are keen on saving the rarest of the bloodlines.

It pays to know your local market and do some homework. For example, we could sell our Middle White fatteners many times over in our area, but a couple of counties away I was told they don't sell as easily as other breeds, possibly as there are more breeders of them in that area. Additionally, some breeds are reported to require less feed, especially when lactating a litter of piglets, e.g. Berkshire; or have higher kill-out percentage, e.g. Middle White; or get to pork weights earlier, e.g. most modern breeds.

Feed is probably the largest cost for most small-scale pig keepers, as labour and building costs are rarely calculated. The same quality feeds can vary enormously in price. As touched on earlier, the ability to buy in bulk gives greater flexibility of buying feed at a lower price but you must be able to use the entire amounts within the stated shelf life, usually around three months, and have suitable vermin-free storage areas.

Comparing the cost of my delivered pig feed (in bags) to what I would pay at my local agricultural merchants, I save well over £100 per tonne, which for us is an instant saving of just over £200 per month. I could save another £100 per month if I had a feed hopper and could buy it loose – also known as having it 'blown in'. To show how researching feed prices can lead to significant savings, my son, Oliver, has made a spreadsheet calculator whereby I input the cost of the feed plus other expenses such as deworming, straw used, vaccinations, plus information such as gilt age of first litter, months empty between pregnancies, weaning ages, number of sows served per boar, etc. and I can see directly how much profit I make per piglet reared. Tables 12.1 and 12.2 show the difference in profit, per piglet reared, between first-time gilts and sows at two

TABLE 12.1 *Feed cost variable £230 per tonne.*

PROFIT PER PIGLET IN FIRST LITTER				SELLING PRICE OF A PIGLET		£50
NUMBER OF PIGLETS	1	2	3	4	5	6
COST PER PIGLET	£327.06	£163.53	£109.02	£81.76	£65.41	£54.51
PROFIT PER PIGLET	−£277.06	−£113.53	−£59.02	−£31.76	−£15.41	−£4.51
NUMBER OF PIGLETS	7	8	9	10	11	12
COST PER PIGLET	£47.63	£42.45	£38.42	£35.22	£32.58	£30.38
PROFIT PER PIGLET	£2.37	£7.55	£11.58	£14.78	£17.42	£19.62
PROFIT PER PIGLET IN SUBSEQUENT LITTERS				SELLING PRICE OF A PIGLET		£50
NUMBER OF PIGLETS	1	2	3	4	5	6
COST PER PIGLET	£155.89	£78.98	£52.66	£39.49	£31.59	£26.33
PROFIT PER PIGLET	−£105.89	−£28.98	−£2.66	£10.51	£18.41	£23.67
NUMBER OF PIGLETS	7	8	9	10	11	12
COST PER PIGLET	£23.47	£21.32	£19.64	£18.31	£17.21	£16.29
PROFIT PER PIGLET	£26.53	£28.68	£30.36	£31.69	£32.79	£33.71

TABLE 12.2 *Feed cost variable £330 per tonne.*

PROFIT PER PIGLET IN FIRST LITTER				SELLING PRICE OF A PIGLET £50		
NUMBER OF PIGLETS	1	2	3	4	5	6
COST PER PIGLET	£436.65	£218.32	£145.55	£109.16	£87.33	£72.77
PROFIT PER PIGLET	−£386.65	−£168.32	−£95.55	−£59.16	−£37.33	−£22.77
NUMBER OF PIGLETS	7	8	9	10	11	12
COST PER PIGLET	£63.68	£56.83	£51.51	£47.27	£43.78	£40.88
PROFIT PER PIGLET	−£13.68	−£6.83	−£1.51	£2.73	£6.22	£9.12
PROFIT PER PIGLET IN SUBSEQUENT LITTERS				SELLING PRICE OF A PIGLET £50		
NUMBER OF PIGLETS	1	2	3	4	5	6
COST PER PIGLET	£199.21	£100.64	£67.10	£50.32	£40.26	£33.55
PROFIT PER PIGLET	−£149.21	−£50.64	−£17.10	−£0.32	£9.74	£16.45
NUMBER OF PIGLETS	7	8	9	10	11	12
COST PER PIGLET	£30.06	£27.41	£25.36	£23.73	£22.39	£21.26
PROFIT PER PIGLET	£19.94	£22.59	£24.64	£26.27	£27.61	£28.74

different feed prices when all the other costs and parameters remain the same – surprising, isn't it?

If you only have a couple of sows, then consider the extra expense of keeping a boar. A working boar eats around one tonne of feed per year (+/– depending upon breed) from around five months of age, and his costs, plus extra straw, wormers, etc., would only be divided by the few litters per year. So it might be worth considering the use of artificial insemination or paying a modest amount for the use of a hired boar. Hiring boars is getting harder as people understandably worry about biosecurity, but there are still a fair few available, especially within the traditional breeds.

You can buy secondhand arks, feeders, water troughs, etc., which would help keep your start-up or herd expansion costs down – but just make sure that you disinfect thoroughly before allowing your pigs anywhere near them. Maintaining good biosecurity including the liberal use of Defra/federal-approved disinfectants, isolating all new or returning stock and incorporating all new stock into your deworming and vaccination programmes will keep your veterinary expenses to a minimum. Here prevention is cheaper than cure.

Creep feeding your piglets can add valuable muscle at an early age. Muscle is built exponentially, so laying it down early can mean slightly less feed later on or reaching pork weights at an earlier age, saving money. Of course if you are selling the pigs as fattening weaners at eight to ten weeks, then you won't see the benefit in terms of meat but you may well get returning customers happy with their well-grown purchases. When we sell carcasses to a butcher, we feed our traditional British Saddlebacks slightly less than the breeding stock to provide a leaner meat, which saves money and often the butcher will pay a better price per kg.

In the UK, don't forget to promote your pedigree meat certificates (Figure 12.1), proving the provenance of your pigs. The BPA has recently introduced the ability to transfer birth notified fattening weaners, at no additional cost, to new purchasers, and if they

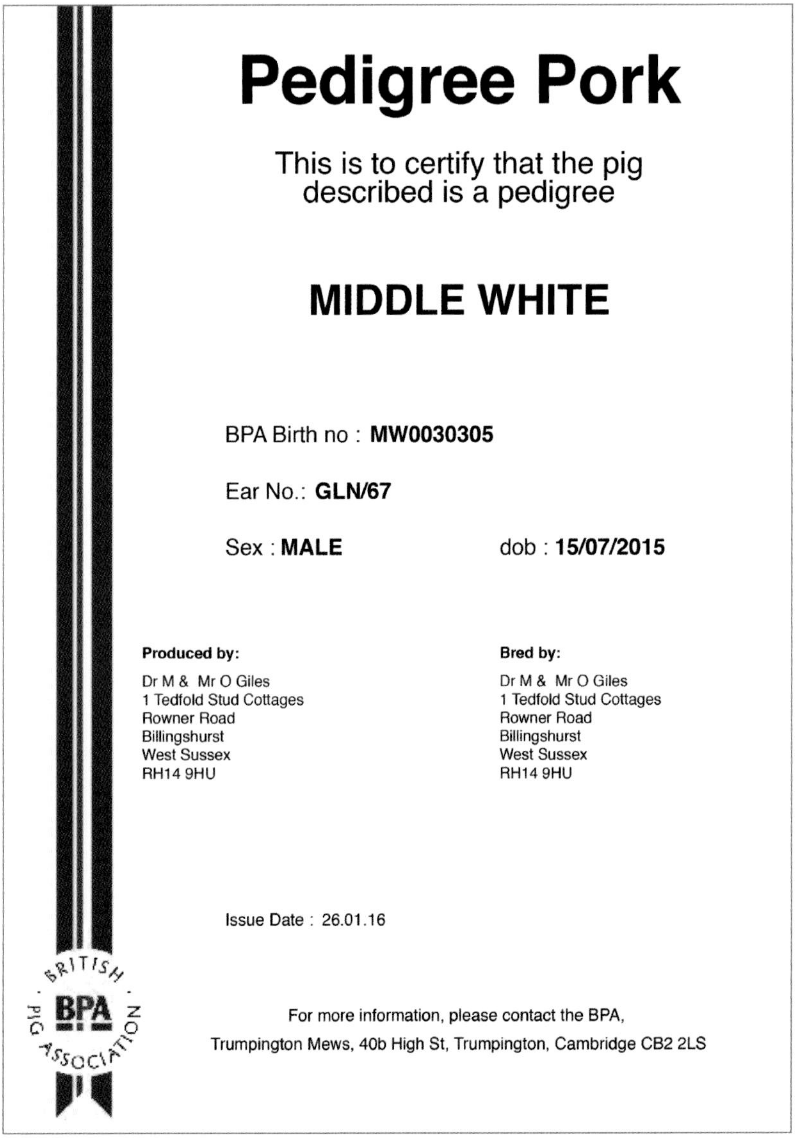

Pedigree Pork

This is to certify that the pig described is a pedigree

MIDDLE WHITE

BPA Birth no : **MW0030305**

Ear No.: **GLN/67**

Sex : **MALE** dob : **15/07/2015**

Produced by:
Dr M & Mr O Giles
1 Tedfold Stud Cottages
Rowner Road
Billingshurst
West Sussex
RH14 9HU

Bred by:
Dr M & Mr O Giles
1 Tedfold Stud Cottages
Rowner Road
Billingshurst
West Sussex
RH14 9HU

Issue Date : 26.01.16

BRITISH PIG ASSOCIATION BPA

For more information, please contact the BPA,
Trumpington Mews, 40b High St, Trumpington, Cambridge CB2 2LS

FIGURE 12.1 A BPA pedigree meat certificate.

are BPA pork producer members they can print off their own certificates when the time comes, as bred by *you* and produced by *them*. Pork Producer membership is currently (2016) £25/year and includes a subscription to *Practical Pigs* magazine. I have already found this really adds to the customer's sense that they really are buying something special – which of course they are!

This next point will be controversial but personally I think it's false economy to start breeding from your traditional breed gilts too early, especially if they are under 130 kg live weight. I know breeders who regularly put gilts into pig to farrow around their first birthday. I'm not passing any judgement but I don't put them anywhere near a boar until at least their first birthday, as I think you get a better-grown, longer, healthier sow later on and the increased number of piglets across the sow's life more than makes up for waiting four more months for the first litter.

However, producing all these high-quality fattening weaners and/or prize-winning-level breeding stock is pointless if no one knows! So some form of advertising is required: this can be through your own personal website, the BPA or TLC website if you birth notify every eligible litter born, through your breed clubs or breed registries' websites if you are a member, through free farming websites, agricultural merchants' advertising boards, attending pig shows with a prominent banner above your pigs, and, of course, social media sites.

Finally, when you do get enquiries, it makes sense to make the inevitable pre-purchase farm tour appealing to the buyer. There is nothing worse than walking past piles of rubbish, redundant machinery, dog muck, ramshackle pens and housing while the seller is convincing you how lovely and healthy their weaners are.

You can now stand back and watch your hard work pay off and let your high-quality, healthy pigs sell themselves.

WHEN TO ADVERTISE

Some breeders advertise their spare stock at a couple of weeks old and people visit and reserve the weaners they want with a small deposit. I would recommend taking a non-refundable deposit, as the number of times in the early days we reserved piglets and then the purchaser let us down was shocking. At least if you have taken a deposit and they let you down, it pays for the piglets' keep until they do sell. Other breeders advertise when the piglets are ready to go; they can be taken the same day and as long as they are not on a 20-day standstill in the UK. There are pros and cons both ways.

WORDING YOUR ADVERTS

You will need to mention the breed or cross-breed in full, the date they will be ready to go, number of males and females available, plus the price and contact details.

For registered stock, you will need to add in the bloodlines for both male and female lines.

For example:

British Saddleback fattening weaners for sale. Ready to go on 12th August. 5 boars and 8 gilts available. £50 each includes Pedigree Meat Certificates. Tel. 123 456 789/email pigs@pigsRus.com

Or:

British Saddleback Pedigree Breeding Stock from Molly/Consort bloodlines. Ready from 10 weeks old on 12th August. Boars from £70 and gilts from £80 includes full registration and transfer. Visitors welcome. Call 123 456 789, email pigs@pigsRus.com or visit website for further details: www.pigsRus.com.

By UK law you can only call a pig pedigree or pure-bred if the parents are both registered pigs and you have birth notified the litter as a minimum. You can only call your pigs by their breed name if they fulfil those requirements, e.g. if you have unregistered Gloucestershire Old Spots that you have bred from, the piglets cannot be named as such in the advert, and are just spotty pigs. If the sire was a registered GOS boar, then you could call the GOS cross-breed.

TAKING CALLS AND VISITORS

Make sure you know how many of each sex and breed you have available or you may sell the same one twice. We have a diary with the number of each sex of piglet and a brief description for each piglet. If the person takes the piglets there and then, we cross the piglets off the list. If they leave a deposit to collect the piglet later, we write their name, and if a deposit was paid we issue a receipt and put their contact details next to the piglet's. If the piglet has no distinguishing features, then we put a semi-permanent mark on the piglet and note that next to the name as well.

When people are visiting to choose their piglets, you have to make time for them. Experienced people will come in, choose the piglets and go within ten minutes. If they have bought from us before, they sometimes tell us what they want over the phone. Less experienced people can take an hour or more, ask a ton of questions and then take their time selecting, as they have decisions to make that they had not thought of, such as what sex to pick. If they are nervous about the whole idea, then offer to let them go away and decide rationally, without the temptation of 20 cute piglets looking at them (Figure 12.2). Obviously, if they are sold before they decide, they may miss out this time, but that is not always a bad thing. Don't hard sell your piglets. Interested

FIGURE 12.2 A pen of temptingly cute piglets.

parties may well buy this time, but they are unlikely to be repeat customers or spread the word.

NOTE: What some purchasers do is treat the experience as a day out and often turn up with Granny, all their kids and a neighbour in tow, all of whom ask questions. Fix that smile on your face firmly in this situation.

We like to deworm our piglets in front of the purchaser at the time of collection, although this is not always possible, especially if we are delivering them. We always write down what the wormer was, and the dose and withdrawal period, and give it to the purchaser. For first-time purchasers we also provide a weaner care sheet with basic feeding and care details plus our phone numbers and email address should they need to ask a question or two.

WHEN THINGS GET TRICKY AND OTHER MISCELLANY

WHEN YOUR PIG IS ILL OR INJURED

There are many factors that will affect your decision about how to deal with a situation, but if you feel that what has happened is out of your league, then contact your veterinary surgeon. Go through the list in the pig health chapter to try and identify the cause, and if it doesn't look too urgent, call or email your mentor or pig course provider to see what they would do, describing what you can see in detail, including temperature and respiration rate if applicable. If you are still not comfortable with the situation, then call your vet. If you have decided that your pig needs to be humanely destroyed, most fell-men can shoot your pig and remove the carcass for incineration.

Mentor and pig-course-provider advice does not replace qualified veterinary advice in any way but new pig owners really do panic sometimes, which is understandable if they care for their pigs.

MY SOW HAS DIED AFTER FARROWING – WHAT SHOULD I DO?

Firstly, let the piglets suckle from the dead sow before *rigor mortis* sets in; you may have to position her so the piglets have full access to the teats. They will still be able to obtain colostrum for a while even without the sow's let-down reflex. If the sow is still warm, then the piglets will get some of their warmth from her, but if she cools rapidly, then a heat lamp will need to be used.

Squeeze the sow's teats to make sure there is still milk accessible, and in any event after an hour or two remove the piglets to a warm environment, e.g. a large box under a heat lamp. Do not have the heat lamp hanging too low or you will burn them. I would also milk the dead sow and use this to give the piglets their precious colostrum.

TIP

Have the emergency box and lamp ready to set up as a contingency plan so you can call your dedicated hand-rearing person (see next paragraph) to come and collect the piglets.

If you have another sow that has recently farrowed (in the past three days), then you may be lucky and be able to foster them on her, but it really depends upon numbers and the attitude of the sow. Realistically the sow can cope with as many piglets as she has teats. Make the orphan piglets smell as much like her own as possible by rubbing on her own afterbirth if you still have it or at least straw from her ark. If you can, mark the orphan piglets so that you know which piglet is from which litter. This is important if they are pedigree piglets and you will be birth notifying them.

HAND-REARING PIGLETS

Unless you can take at least three weeks annual leave at short notice, then this activity is not possible for the *Commuter Pig Keeper*. Luckily it is not that common an occurrence – in all the years we have been breeding we have had one piglet that was rejected and one litter that required milk due to the sow losing her milk supply because of a metritis. But it is your responsibility to know and have a contingency plan for what you are going to do with orphan or rejected piglets, even if that is to ask a vet to kindly put them to sleep. Do not abandon, drown or starve them – it is not only illegal, but also barbaric.

We have a couple of people who are prepared to step in and hand-rear for us. They know what to do and we would give them all the necessary equipment, which we have ready in a box.

You can buy commercial sows' milk powder but it is expensive and the shelf life is not always that long, so we have ordinary human baby milk powder as standby, which is made up exactly as it states on the packet. If it is obvious that the piglets are going to survive, I would then go and purchase sows' milk powder.

To begin with, use a small baby's milk bottle with a medium-flow teat and offer ad lib milk every two hours. They will probably take only a few millilitres at a time to begin with, but persevere; as they get stronger, they will take increasing quantities. After a few days, start to leave slightly longer periods between offering the milk, as this is often a good way of making them take more at each sitting. They are not very cooperative to begin with and wiggle constantly, so I sit on the floor with a towel over my lap and rest the piglets across my thigh, keeping their back legs touching the floor. Other breeders start them off in a shallow tray from day one and persevere to encourage them to feed by putting them at the tray and gently dipping their snouts and hoping they get the idea. If you find a method that works for you, use it.

The piglets will definitely require an iron injection of 1 ml into the thigh muscle, so if you do not have any, then order some straight away or ask your vet to do it. When the piglets are seven to ten days old, start introducing creep feed, and if you have used the bottle method, now try to get them to drink their milk from a tray/bowl so you have less of a chore. You will also need to provide access to water from aged two weeks.

With good luck and perseverance you may be lucky and rear the litter, but you should know that hand-rearing piglets is not always successful.

MY PIG IS DIGGING TOO MUCH

In truth it's kind of tough! It's what pigs do, so if you cannot accept it, then pigs are not for you. Some breeds naturally root less, preferring to graze if there is enough grass in the pen, e.g. Kunekune, Large Black, so perhaps try these breeds. You can put rings in their noses, but I'm afraid you will not read how to do it in this book as I consider it an unnecessary mutilation and an infringement of their piggy rights.

MY PIGS HAVE EATEN MY CHICKENS

Pigs are omnivores and are only fed a vegetarian diet for biosecurity reasons. Given the chance, pigs will kill and eat chickens (or other animals) that venture into their pen and are slow enough to catch. Most often it is chickens that have remained inside the ark/pen at night-time and tried to roost there.

MY PIGS ARE EATING STONES

Some pig keepers say that the pigs are cleaning their teeth when they roll stones around in their mouth, while others say it's a boredom thing and the pig's equivalent of chewing gum. Either way, some pigs do it while others don't, and we have had a couple of sows over the years that did it regularly. I sometimes think it might be a way of creating more saliva, which may help with indigestion, but that is just a guess. If the pig swallows lots of stones, it may cause a digestive problem and while I have heard of this happening, it is not a problem I have come across.

MY PIGS DANCE ON THEIR FEET WHILE EATING

If it is occasional and they are moving about to keep a comfy position, i.e. no more than you would move a foot when standing still, then it is normal. If they are hopping about from one foot to the other, there is a problem.

In growing pigs that do it with their hind legs it could be a sign of osteochondrosis – a type of arthritis – and you may just have to hope that you get them to pork weight; if it's in young breeding stock, then you will have to consider their future carefully. Before culling, check there isn't something obvious with their feet causing it. We had an issue with a couple of sows who when they farrowed inside seemed to 'go off their feet' and had been known to crawl to their feed bowls. We eventually worked it out and it was their cleys becoming sore from a combination of the rougher concrete surface and when they walked in their urine; it always cleared up when they were put back outside. Another good reason why we farrow inside only when there is no other choice.

MY PIGS SPEND LOTS OF TIME SLEEPING – ARE THEY OK?

More than likely they are fine. Pigs are almost crepuscular by nature, which means they are most active at dawn and dusk, which is strange really as their eyesight is atrocious and you would have thought they would need full light. I say 'almost' because if we went up at dawn to feed ours, then we would have to wake up every single one of them; but older pigs, especially, get up in the morning for breakfast, have a rummage about and then go back to bed until nearer evening feed time. Younger pigs can be seen during the day digging and rooting about more often, and all pigs are happy to get up for a chat or treat whenever they hear you. If they don't get up to eat, then that is the time to go through your 'is my pig ill?' list.

MY PIGS INSIST ON SLEEPING OUTSIDE

It is annoying when you have purchased an expensive ark and they don't sleep in it, especially if they drag all their bedding out to do it. In the summer months it's a pretty normal occurrence and they enjoy sleeping outside (Figure 13.1). If they are doing it in the rain or colder months, it isn't normal. In all cases, check the ark isn't touching the electric fence or the bedding is mouldy or dusty and rectify any problems you can see. Also check they haven't been using the ark as a toilet and the ammonia levels are high.

FIGURE 13.1 Pigs enjoying sleeping outside.

MY SOWS WERE FRIENDS BEFORE AND NOW ARE FIGHTING

When pecking orders in a herd are broken by a prolonged absence of one or more of the sows, then it needs to be re-established. The more dominant sow will make sure the other sow isn't going to challenge her authority, by being overly aggressive. If there hasn't been an absence, then it is likely to be less aggressive and just a natural rejiggling of the pecking order.

We learned the hard way to the detriment of one of our sows. The two pigs concerned had been brought up together quite happily, but were then separated for farrowing and were apart for eight weeks. Once the eight weeks were up, we just put them back together, whereupon one sow charged and took out the other one, causing severe muscular hind leg damage. By rights she should have been put down, such was the extent of the injury, but as she was a very rare bloodline, we were keen to see if we could save her. It took months, during which time she could not be bred from or produce piglets. We also didn't know if she could ever take the weight of the boar again, we worried she might go barren during her pregnancy-free months and we did not know whether she would be proficient enough during farrowing not to squash every piglet when she laid down.

FIGURE 13.2 Pigs being fed over a wide area.

Although she still walks with a slight limp, she can take the weight of smaller boars, and has successfully raised a few litters of piglets since, but she will be prone to arthritis as she gets older and cannot be sold on. Luckily she is a sweetie to look after.

When introducing sows, we now separately pen them side-by-side and feed them with the electric fence in between them for a week or two before introducing them to one pen. We try and do the mixing just before dark and feed over a wide area (Figure 13.2). This seems to keep the fighting to a minimum and after a night's sleep together they start to become friends.

BOARS FIGHTING

Entire boar weaners can be kept together until slaughter weight quite happily if they are brought up together; their play can be a little boisterous, but it is only play. Young boars that have started working and/or older boars must never be mixed together. They will fight to the death, without a shadow of a doubt.

If they do accidentally get in together, the fight will start immediately and it isn't going to be an easy task to separate them. If you are going to be successful and not sustain an injury yourself, then wading in with a stick and a board will not be enough. You can try throwing some food into the pen to aid distraction but don't expect it to work. If it does, then get the easiest/nearest boar to the gate to move out and put it somewhere safe.

The safest option by far is to put a few, preferably in-season, sows in the pen, as sex is probably the only distraction that would make them stop fighting – and then while one of the boars is 'busy', remove the nearest boar. However, if you will be in any danger at all, leave them to it and prepare to deal with the ramifications.

TAIL, EAR, FLANK AND VULVA BITING

Outdoor-reared pigs rarely, if ever, exhibit vices such as these. I would take a good hard look at your husbandry and the stocking rates in your pens if this is happening. Give the pigs added interest in their pens when the ground is baked hard, such as a few heavy logs (heavy ones because otherwise they will push them onto the electric fencing and short it out), clumps of freshly dug earth, or piles of hay to chew.

Minor ear biting may occur in young piglets during playing or when you mix litters together to establish a hierarchy, but it is rarely more than a scratch then. We sometimes see bloody ears on piglets when they are being exhibited at shows, especially when it is a three-day show and the pen is not very big. If you do get a chewed part of an anatomy, then treat as a wound to avoid infections.

MY GILT/SOW/BOAR KEEPS RUSHING AT ME AND PUSHING ME

If this activity is only at feed time and at all other times they are OK, then I would just throw some feed in first to distract them to give you time to put the food where you want to. Or choose a place where you can just throw the feed from the fence line – which is easier for the *Commuter Pig Keeper* anyway.

If it is a very young pig, treat it almost as you would a puppy. Give a loud 'NO' followed by a rap on the snout with your hand (once) and stop treating it to scratches and belly rubs for a few minutes by pushing it away. Repeat as necessary, but hopefully after a few goes it will be behaviour of the past.

If it is only when your sow or gilt has a litter of pigs, then this can be considered normal with no discipline required. Although you may wish to sell her on before the next litter, tell the new owners that it occurs in case they too are *Commuter Pig Keepers*!

NOTE: If you have Mangalitzas, Iron Age pigs or Durocs, this is super normal as they are fiercely protective of their babies and I did warn you earlier in the book.

If it occurs at any time and is just part of their character, then for working boars it is sausages, I'm afraid. There is no room for boars with this nature for the smallholder, let alone the *Commuter Pig Keeper*.

If it is sows or gilts, you may stand a chance of correcting the behaviour with firm discipline. Arm yourself with a board and stick – the board is to protect you, and when the sow/gilt rushes at you, a well-aimed sharp smack across the snout with the curved end of the stick should stop her in her tracks. You may have to repeat a few times, but never beat your pig or she will turn vicious, understandably. Do reward her if she stops in her tracks. If she stands there looking at you and not attacking, then end that session on a good note with a few sow rolls or some grapes. Hopefully the next time she will think twice, but repeat if she doesn't. Don't be weak about it or expect it to work first time. If after a few times she is still exactly the same with no improvement seen, then it's sausages, as there is no time for this in your day.

If the pig is downright vicious in the chase, teeth out and determined to get you, then you can try the above technique, but your chances of success are slim, so make that call to the abattoir or fell-man.

MY PET PIGS YELL EVERY TIME THEY SEE ME

It's unlikely when there is more than one pig that they are lonely or craving human company. It's more likely that you give them food or a treat every time you pop to see them, so when you do happen to walk past, they expect to be fed. Keep treats restricted to meal times and go and see them without treats until they used to the idea. If you only have one pig, then not only does this contravene the Five Freedoms (see Chapter Two), but it's not fair and you need to get another one for company.

MY NEW WEANERS WON'T LEAVE THE ARK

Scared piglets will retreat to their ark and if they are not very brave will only venture out when you have gone; piglets that have not been socialised with humans much and those who have had a bad experience when being put in the new pen, e.g. being bitten by electric fencing, may be agoraphobic in the beginning.

You can put the feed and water outside and force them to come out, although to start with I would put both water and feed close to the doorway so a) they know it is there and b) they can at least have a drink. They will take some comfort that all their amenities are close enough to the safe environment of the ark.

As they get braver and venture out, then move the feed and water nearer to where they suit you. Try going into the ark slowly and with gentle handling and the cunning use of homegrown grapes or cherry tomatoes let them know that you are lovely. Don't terrify them, just sit in the doorway to start with, then quietly reach in and touch them, etc.

I DON'T HAVE ACCESS TO A TRAILER

If you can't beg or borrow a trailer, there are small companies that you can hire to take your pigs from farm to farm/farm to abattoir for a fee, usually a fixed call-out plus mileage. While this will add to your total costs, it is cheaper than buying a trailer if you only have a few pigs to send off once a year. Some abattoirs also offer a collection service, again for a fee. Your pigs will need to be contained prior to collection in a small area accessible to the trailer as the haulier will not chase pigs around a paddock or field in order to load them.

DISEASE OUTBREAKS

Disease outbreaks may stop you moving your pigs anywhere, including to slaughter sometimes. You will be notified by various media, primarily the television and radio. You may not receive a phone call, text, email or visit to tell you, so keep the radio on in the car. You can register with state and government departments to receive text and email alerts, and this is worth considering, although these days Twitter or Facebook are likely to provide your first alert.

Abide by what you are told until you hear otherwise, or face a fine or imprisonment. The security of the national pig herds, which includes your pigs, is of prime importance. Worth noting is that if your pigs are registered with the Rare Breed Survival Trust – and if you own BPA-registered pigs, this happens by default – then the utmost will be done to keep your pigs from compulsory slaughter. It isn't always possible, especially if you are in the epicentre, but it is another good reason to be picky about sticking to your biosecurity plan.

UK list of notifiable diseases: www.defra.gov.uk/animal-diseases/notifiable
USA list of notifiable diseases: www.aphis.usda.gov/wps/portal/aphis/ourfocus/animalhealth
Worldwide list of notifiable diseases: www.oie.int/animal-health-in-the-world/oie-listed-diseases-2015

SLAUGHTERHOUSES ARE SHUT AT THE WEEKEND

Most, but not all, abattoirs are shut at the weekends, so it's likely you will have to arrange all your slaughter dates around your work commitments or hire one of the small companies/professional hauliers to transport your pigs for you. Most hauliers would collect your pigs quite early, which should hopefully give you enough time to get to work. Some abattoirs will allow you to drop them off the evening before, but personally I would prefer not to, as it will be a strange environment for them.

OFFICIAL GOVERNMENT VISITS

These will all occur during the week, as officials do not work at the weekends except during disease outbreaks. They have the right under EU or USDA law to only give you 24 hours' notice of a routine visit and can report you as non-compliant if you refuse.

Having said that, they do try their best to accommodate the smallholder, but you may have to do a bit of fast explaining to your boss to get some annual leave or arrange an early leaving/late start on a particular day.

In the UK, it is highly likely you will get a Trading Standards visit in the first month or two of keeping pigs to check you know what you are doing, and again they are as flexible as they can be. It is also a possibility that a sneaky neighbour, horrified at the thought of having pigs nearby, might report you to an official body with false allegations, and again they have to visit to check all is well. It is worth involving the neighbours with the promise of some delicious sausages in a few months' time if at all possible, but there is 'nowt as queer as folk' so don't be surprised if they are initially resistant.

MY KIDS HAVE BECOME INTERESTED AND WANT TO KNOW MORE

Excellent! Apart from helping you in the day-to-day chores, there are also official clubs that are exceptional in not only engaging young people but also educating them along

FIGURE 13.3 A winning pig-and-child combination in the show ring.

Photo courtesy of Miss Jolie Matthews of the Lowpark Herd.

the way. In the UK, the National Federation of Young Farmers' Clubs, which promotes farming and rural activities, has over 600 regional clubs for young people aged 10 to 26 years old. The British Pig Association has a Junior Pig Club with educational pig events such as sausage making, artificial insemination courses, carcass competitions plus regional pig-handling prizes at most shows, culminating in a championship stockman and handler competition (Figure 13.3).

In the USA, the 4-H Positive Youth Development Program includes farming and showing and has clubs, camps, afterschool and school enrichment schemes in every county and parish.

The National Future Farmers of America Organization (FFA) makes a positive difference in the lives of young people by developing their potential for premier leadership, personal growth and career success through agriculture education, and the activities are all linked to middle and high school students wishing to pursue a career in agriculture.

For children specifically interested in pigs, the National Junior Swine Association linked to the National Swine Registry hosts many junior events around the country, and college scholarships are even awarded for those who show great promise. Team Purebred – the Junior Purebred Swine Association – is another organisation with a mission to provide educational and career opportunities through competition and fellowship in order to develop leadership skills and integrity in young people interested in Berkshire, Chester White, Poland China and Spotted Swine.

THE 'OTHER HALF' INSISTS ON A HOLIDAY

If you are a man, then I can recommend diamonds as a pacifier; if you are a woman, then you will already know how to get your own way! Seriously, if you only have a few fattening weaners a year, then time their arrival so that they will be in the freezer when you want to go on holiday, or so you collect them after you get back. If you have breeding sows or pet pigs, then you will have to arrange some cover from someone who knows what they are doing.

WORK DEMANDS YOU ARE TO ATTEND A THREE-DAY COURSE

Annoying, I know, especially if it is one of those horrid team-building ones! If you have a partner in the venture, then so much easier and you can offer to do both morning slots at the weekend. Who would say no to that? You will need to arrange some decent cover – it really is as simple as that.

PREDATORS OR FERAL WILD BOAR

It is your responsibility to protect your livestock from being attacked and it makes biosecurity and disease prevention sense as well.

Double fencing combined with electric will assist in keeping some predators out but not all. In the UK, there aren't many predators apart from domestic dogs, or in some locations feral pigs that choose to attack pigs of weaned age and above, and so locking them away at night isn't required. The only time we have experienced an attack was when the sow was actually giving birth and a hungry fox or badger took its chance. The birthing fluid odour definitely acts as an attractant, as our farm dogs know when a birth has taken place and always rush, sniffing at the air, to the appropriate pen.

In the USA, there are many species of animal that will tackle any sized pig, including cougars, bears, mountain lions, etc., and so the husbandry needs to reflect the level of local danger, with perhaps robust housing at night as a minimum. The use of Livestock Guardian Dogs (LGDs) seems to be a popular option in the USA with some livestock keepers, reliably alerting owners and even fearlessly chasing off the predators. My American reviewer of this book, Christopher Rowley of Goose Meadow Kunekunes, told me that he has four Karakachans (LGDs) on his property and has never had coyotes, foxes, bobcats, feral hogs or any other predator crossing his fence line.

MORE FUN YOU CAN HAVE WITH PIGS – SHOWING

Before I go into what we know about showing pigs, apart from that it's great fun, I must point out that we have only been showing since the beginning of 2010 and are rightly considered 'newbies' on the show circuit. We probably still will be in 2029 as well because neither of us was born into a pig-showing family, but the teasing we get is all good natured.

However, we have taken every opportunity to learn as much as we can and literally hang on every word of the experienced showmen. The showing advice in this chapter cannot be taken as a definitive – we are in no way experienced enough to claim that we could achieve even near that goal – but it will provide a good place to start. I would also be the first in the queue to buy the definitive pig-showing guide, should one of the experienced showmen or women decide to write one – hint!

TOP TIPS ON HOW TO KEEP YOUR EXHIBITION STOCK IN TIPTOP CONDITION BEFORE AND DURING EVENTS

What's the difference between show condition and healthy condition? The physical condition of the pig should be remarkably similar to those sleeping in the arks at home, especially the existing and future breeding stock.

Keep your show pigs at a nice body condition score of 3 to 3.5 (see Chapter Three). You don't often see scores below 3 on the show circuit and the general trend is for pigs to be too fat. A little over can be tolerated, as travelling often brings female pigs into season and some ladies can get a little tetchy and slow up on their eating during this time, and the boars upon arrival can be all consumed mentally in smelling those lovely ladies – so a few extra kilograms can see them through still looking good.

I wrote an article about showing a few years ago for the BPA *Practical Pigs* magazine, and after it was published, I had a fellow stockman approach me, saying that I didn't follow my own advice and my gilts were a tad fat. He wasn't wrong, as a couple of them really were, and I did try and rectify it before the show season. But it's an awful lot easier not to let a pig get too fat in the first place, rather than try to put it on a diet.

It was obvious with our gilts what the problem was – the only fat ones were my son's favourites, and all those extra handfuls of sow rolls at feed time had mounted up!

It's good practice to deworm and vaccinate before the show season starts. Vaccination decisions can be tricky as you have no idea what could be passed around that year. I tend to concentrate on the ones that could affect a breeding herd, such as *Erysipelas*, but decide in conjunction with your vet. In the USA, some states may have mandatory vaccinations.

Pigs in good health are more resilient to pathogenic bugs so don't panic but do remain observant. In the USA, you can buy 'show chow' – a specific feed for show pigs, which contains antibiotics to protect the pig against mostly respiratory *Mycoplasma* infections. Personally I think providing antibiotics prophylactically should be prohibited as it's the quickest way to select for antimicrobial resistant bugs and contributes to global antibiotic resistance. Save antibiotic use for when a pig is actually unwell.

Worm eggs are sticky chaps, so deworm your pigs before the show season starts with a dewormer that is known to be effective in killing worm eggs, so not an ivermectin-based one, and repeat before putting the show team back in with your other pigs at the end of the show season. We usually wash our pigs' legs down with Hibiscrub™ or liquid soap after each show to help remove any eggs stuck to their hair, which could end up ingested; it also helps in removing any pathogenic bacteria that they may have trodden or laid in. It is also important that if your pigs succumb to an illness before the show, you don't take them. Not only are they more susceptible to catching something else, but they may also be infectious to other pigs.

Keep your pigs fit. If they are not used to travelling and/or walking around for any length of time, then they could find it quite exhausting being in the ring for different classes or if you're lucky the championships. Some of the larger class numbers seem to go on forever and your pig has to be fit enough to still look fresh at the end of the class. This is more important if they are kept indoors and don't have enough space to run and walk about. We use a steepish hill on our land to assist in muscle development when we practise our stick and boarding, although our pigs, being outdoor kept on heavy clay through the winter, seem to find it easier than the handlers do.

Good skin can make all the difference to the pig's health and the colour of the rosette you receive. So provide plenty of shade and wallows for outside-kept pigs; you can't scrub away sunburned skin. I've seen the odd white pig in the ring with red crusty sunburn, and not only is it painful for the pig, but it can take away a winning position and reflect upon the care of the owner. Some pigs' skin can get quite encrusted with dirt over the winter, or they may have stained legs from the winter mud, so pick a nice day before the shows start to give them a good wash with shampoo and soften their skin with some pig oil. If they are really scurfy, then a mix of pig oil and sulphur can remove it effectively. The preparation applied after the last bath before the show ring will depend upon what breed you have; the white breeds have a wood flour applied to assist in cleaning the coat, the theory being that the wood flour mops up any dirty grease on the hair and skin; coloured breeds are usually oiled to a nice sheen; but Tamworths and Mangalitzas enter the ring *au naturel* (Table 14.1).

Go easy on the use of oil if it's a hot day as this may just heat up the pigs to the point of heatstroke. On those days, tepid water applied just before entering the ring assists in keeping the pigs cool through evaporation. It can even be applied through the day in their pens to help keep them comfortable.

Keeping the pigs as stress free as possible is an important requirement. Getting them used to loading on to the trailer followed by a pleasant journey to the show with weather-appropriate ventilation provided is a must. If it's a really hot day, then travelling earlier or later in the day to avoid the extreme heat helps keep the stress levels down.

Once at the show, wherever possible choose a pen for the boars that is not near other boars, e.g. next to your kit pen, and remember there may be boars in the attached pen behind. Show folk are usually keen to help juggle their boars about as well to avoid potential boar arguments. A well-known UK showman of British Lops uses a flavoured multi-vitamin in the water at home prior to show season, so when her pigs attend shows and are required to drink strange water, they don't notice.

You would have thought with technological advancement in most areas of life that the preparation of show pigs was much different 60 years ago than it is today. The advice given in 1955 is not much different to what we should do today, as I found out while flicking through an old copy of the *Pig Gazette*. Some things obviously cannot be improved upon.

So keeping your stock in good condition before and during events needs to attend to both the pigs' physical and mental needs. If you make showing an enjoyable occurrence for the pigs, then they'll let you know by happily loading up for the next show. Some of ours even load themselves into the trailer without assistance, they are so keen.

THE BASICS OF SHOWING

Pig Identification

In the UK, there are three types of shows: British Pig Association (BPA) accredited, BPA affiliated and non-BPA shows. Only BPA-registered pigs with the appropriate primary identification may be exhibited at BPA shows. Primary identification requirements (Chapter Seven) are in addition to the Defra ear-tag requirements, with the breeder's

Herd Designation Letters (HDL) also printed on the tag. Affiliated shows additionally accept double-ear-tagged pigs. The British Lop and Kunekune Societies are independent and their identification rules must be followed. Non-BPA shows only need the Defra requirements of one tag with your herd mark/number on.

In the USA, there are many events that have pig showing, and may be named Extravaganza, Expo, conferences, fairs or shows. The National Swine Registry and National Junior Swine Association hosts 19 events around the USA for their pig breeds, and pigs must be identified, DNA stress gene tested, and registered prior to entry, plus possess all interstate movement health checks and licences to be able to exhibit. With the exception perhaps of the Kunekune breed, there sadly seems to be a lack of shows that include heritage breeds, and there is only one way to sort that out, heritage breeders of the USA!

Selecting a Pig with Show Potential

To win, the pig must be the best example in the ring. So when selecting a pig for showing, clutch onto a copy of the breed standards until you have acquired a trained eye (Figure 14.1). Look at your pigs without emotion, as cute eyelashes and a double curl in the tail won't cut it with the judge!

There a few things to consider across all breeds. You may not think the judge will get down and check their underline (Figure 14.2), but they will, so check that all teats are evenly spaced; ideally they should mirror each other exactly in a 'domino' pattern down the belly and be in equal number both sides. Blind, dummy and inverted teats are not always tolerated in the show ring, and this includes boars, which require at least three pairs of teats in front of the sheath. The breeds vary in minimum teat numbers, so check the breed standard.

Next, observe your pig when moving; walking badly is the first thing a judge notices, so select a pig with upright pasterns which enables a straight, free walking action with front feet that don't turn in or out (Figure 14.3). The back feet mustn't touch, so check their hocks aren't too widely spaced or close together. Aim for well-muscled hams that are in balance with their shoulders in a Marilyn Monroe-type way when viewed from above, and leave at home any pig that dips its back or slopes sharply at the ham. It's also worth considering the temperament of your pig. If it hates its first outing, why put it through it?

When showing becomes an addiction, the timing of your pigs' farrowings becomes more important and you want to present as well grown a pig as possible for inspection. There are usually classes for older sows, boars/gilts born after 1 July the previous year and after 1 September the previous year, as well as classes for pigs born after 1 January in the current year. Once pigs are fully grown, e.g. in a sow class, then their date of birth becomes less important, but they must have reared a litter in the previous six months or

British Pig Association
Breed Standard and Standards of Excellence
BRITISH SADDLEBACK

Section A - To be eligible for Herd Book entry a pig must (except in exceptional circumstances) be:
- bred in the United Kingdom or the Republic of Ireland
- have at least 12 sound teats
- ear-marked and birth recorded with the BPA in accordance with current regulations
- the offspring of parents already registered either in the Herd Book of the same breed (or in a supplementary register of the Herd Book maintained at BPA's discretion)
- Free from congenital defects (e.g. Umbilical and Scrotal Hernias , Atresia Ani (blind anus), cryptorchid boars, extra cleys, twisted, overshot or undershot jaw and rose on the back)
- conform to any other such regulations as are made by the BPA Council from time to time

In exceptional circumstances a pig which does not fulfil all the criteria above may be accepted for herdbook registration following an inspection.

Breed Specific Requirements for Herdbook Registration

Ears	Carried forward
Colour	Black and white with a continuous belt of white hair encircling the shoulders and fore legs White is permissible on nose, tip of the tail and the hind legs but no higher than the hock.

Breed Specific Disqualifications which make the pig ineligible for Herdbook Registration

Colour	An animal not possessing a continuous belt of white hair over the shoulders and fore legs Chocolate or red coloured White hair on the face or body outside the band over shoulders and fore legs, and tip of tail

Section B – Standard of Excellence – These are recommendations only - breeders should try to achieve these standards in their breeding programmes. Pigs will be judged against these standards of excellence at BPA shows.

Head	Medium length, face very slightly dished, under-jaw clean-cut and free from jowl Medium width between ears.
Ears	Medium size, curbing but not obscuring vision.
Neck	Clean and of medium length
Shoulders	Medium width, free from coarseness, not too deep.
Chest	Wide and not too deep
Back	Long and straight
Ribs	Well sprung
Loin	Broad and strong and free from slackness.
Sides	Long and medium depth.
Hams	Broad, full and well-filled to hocks.
Legs	Strong with good bone, straight, well-set on each corner of the body..
Feet	Strong and of good size.
Coat	Fine, silky and straight.
Underline	Straight; with at least twelve sound, evenly spaced and well-placed teats starting well forward
Action	Firm and free

Breed Specific Objections – Breeders should try to avoid these in their breeding programmes

Ears	Pricked or floppy
Head	Badger face, short or turned up snout.
Skin	Coarse or wrinkled.
Hair	Curly or coarse coat, coarse mane
Teats	Unsound or unevenly placed teats.

Pigs must comply with Section A. Breeders should aspire to breed pigs which meet the Standards of Excellence in Section B. More information is available in the BPA leaflet – Pedigree Breeding the Next Steps.

British Pig Association, Trumpington Mews, 40b High Street, Trumpington, Cambridge, CB2 2LS
Tel: 01223 845100   e-mail: bpa@britishpigs.org
www.britishpigs.org
Updated Autumn 2013

FIGURE 14.1 An example of a breed standard document.

FIGURE 14.2 Judges take seriously identifying the perfect pig (and not how it is now imo)..

FIGURE 14.3 Oliver showing his favourite Middle White sow.

be pregnant. Kunekunes have breeding and pet classes linked to their year of birth but must be over six months of age to enter the ring.

Birth timing will also be a factor in the USA, but the ages and birth date cut-off points differ in range, so check the shows you are likely to exhibit at a year in advance and time your litters born accordingly. There will also be classes for barrow pigs, which may or may not be shown with the boars, depending upon the show and entry numbers.

Showing Boars

Most BPA shows accept entire boars born from the previous year's July onwards, but a few have classes for any-aged senior boars. It's a requirement when showing boars of any age that two competent handlers, both equipped with sticks and boards, are present. One person controls the boar while the other remains alert to the presence of the other boars and makes sure no fights break out. This is also applicable when showing a boar in pairs and groups of three, so that's up to four handlers not blocking the judge's view. Boars over 12 months must also have their tusks cut before the show.

We have shown January-born boars in the ring without hitch or problem from our second year of showing, but we made the mistake of showing a July-born boar before we were ready – although in our defence he was a sweetie at home so we had no idea

FIGURE 14.4 *Two experienced handlers struggling with a July-born boar in the show ring.*

Photo courtesy of Liz Shankland.

However well behaved your
boar is at home, please
don't attempt to show any
aged boar until you have
had lots of experience, and
when you do, start with
those born that year.

of what lay ahead. He was near perfect in looks and conformation and was awarded reserve breed champion on his first time out – but his behaviour was atrocious, even dangerous. He took one whiff of the air full of lovely smells and fought both handlers the whole time, objecting to the board and the stick.

The judge didn't ask us to leave, as the handlers were coping, but it was not a pleasant time for anyone (Figure 14.4). Outside of the ring the judge also gave us some tips for his next outing, to try and stop him smelling the ladies and the other boars, such as menthol rub, e.g. Vicks™, on his snout and on the board and stick, or smelly deodorant on the board. We did this at the next show and it made no difference. He was vile again, so we withdrew him from the third show he was entered into, for everyone's sake. Most of our fellow competitors were supportive and someone we admire greatly said they had a similar time early in their showing career with one of their July-born boars as well.

Finding Shows and Completing Your Entry Form

In the UK, the BPA advertises their accredited, affiliated and 'show and sales' on its website and in its official *Practical Pigs* magazine. There are also websites that advertise various shows around the UK, including 'local' or non-BPA shows. In the USA, the breed registries advertise the shows sponsoring their specific breeds, with hyperlinks to the individual show websites.

Each show will have different methods of entering; request or download a show schedule and send in the entry form and payment by the deadline. Nearer the show date you will be sent any entry passes required and full instructions of what to do.

What Do the Judges Look For?

Judges look for a well-controlled, forward-moving and beautifully presented pig. They will watch the pigs as they enter the ring and often spot the one that has that something extra. Each pig will be closely inspected in turn for points of excellence and for faults as set out in the breed standard. Dispensation on 'well-behaved' is given to January-born pigs, with some allowance made for their speed and exuberance for life at that age (Figure 14.5).

As the judge examines each pig, they will start from the feet up, so if your pig can't walk properly, it doesn't stand a chance. Judges like to flick their eyes between the pigs before making their final selection, so make sure you keep your pig walking well when they are looking. In a ring of high-quality pigs the decision may be based purely on judge's preference; sometimes the decision is controversial, but must be accepted as how that judge saw your pig on that day. If you want a heads-up, then attend shows

FIGURE 14.5 *Dispensation is given for the exuberance of young pigs in the show ring.*

without a pig and watch, perhaps befriending someone for tips, and consider a BPA, NJSA or privately held course on showing and handling.

NOTE: *Sows/gilts halfway through a pregnancy look their best in the show ring and you cannot travel pigs in the last 10% (11.5 days) of pregnancy for obvious reasons, so before you enter her into a class, think ahead if she is pregnant as to when she will be giving birth.*

Preparing the Show Pig

You are aiming for 'fit not fat', so your feeding should be as per usual for the most part. Approximately a month before the show classes, if required slowly increase their feed slightly to give a slightly better covering, to around a body score of 3.5 maximum.

Regular walking for exercise and to encourage muscle development in the shoulders and hams is a must. Start about a month or so before the first show in short blocks of time, using a penned off area if possible, and always finish on a good note with a treat. This exercise should be in conjunction with stick-and-board training rather than following a bucket of food.

The board, used for steering, stopping and blocking, is carried in your left hand, and the stick/paddle used for motivation and acceleration is carried in your right hand. You can buy commercially made paddles that are flat and wide, or you can use a walking stick held upside down with the curved end nearest the pig.

Aim for a clockwise direction and tap around the shoulders to get them moving forwards; tap slowly to maintain the pace and quickly to accelerate. Tapping their hams can make them hunch up, so do so sparingly and never tap the eyes, genitals and joints, as this will obviously hurt. If your pig turns around in your practice sessions, guide it to resume the clockwise direction as soon as possible.

If possible introduce your pig early to bathing and brushing. Preferably use warm water, as not only does it get more dirt off, but it's nicer for your pigs, and use a soft brush or flannel on the face, being careful to avoid their eyes. Always make sure the shampoo is rinsed off properly or you will get scurf forming, and dry off excess water with a towel. For stubborn scurf use pig oil, with or without sulphur added, to soften the skin and make it easier to remove. If using sulphur, make sure it's all removed after use, and don't use oil in hot weather on outdoor pigs or they will cook.

The biggest risk to your herd health is mixing with other pigs, so consider your vaccination programme and the strategic use of isolation pens between shows and before mixing show pigs in with your non-showing stock.

Pre-show Preparation

Create a secure area for your pig near a source of water and use food to keep your pig stress free and occupied (Figure 14.6). Wash your pig using shampoo – a few times might be required if your pig has any white hair and has been living outside in the mud. Pay attention to the insides of the ears as well, especially in prick-eared breeds.

FIGURE 14.6 Scrubbing pigs clean in preparation for the show ring.

Traditionally, how the pig is prepared after bathing depends upon the breed (Table 14.1), and as artificial whiteners or dyes are not permitted, invest in a decent scrubbing brush.

TABLE 14.1 The show preparation method for each pig breed.

BREED OF PIG	PREPARATION METHOD
Tamworth	Bath only
Mangalitza	
Landrace	Wood flour (Figure 14.7)
Large White	(NB: The spots on a GOS may be oiled)
Welsh	
Middle White	
Gloucestershire Old Spots	
British Lop	
Piétrain	Pig oil
British Saddleback	(NB: Only the Piétrain spots are oiled)
Berkshire	
Large Black	
Hampshire	
Duroc	
Oxford, Sandy and Black	
Kunekune	

FIGURE 14.7 Wood flour being used on a Gloucestershire Old Spots.

Print out a copy of your pig's registration certificate to display on its pen. It's a BPA requirement to have a copy with you, so why not display it? Some shows offer prizes to the best-decorated pen, so pull out that bunting and have a banner made promoting your herd.

The morning of the show, sponge wash any dirty areas if required, and rinse. Reapply pig oil/wood flour, remembering to wipe/brush off the excess before entering the ring. Most of the larger shows have an on-site washing facility, which makes life easier.

Showing Etiquette

NOTE: There is no such thing as the perfect pig, and it's the judge's job to find the faults and your job to 'legally' hide them.

You will need to be smartly dressed, so a shirt and tie, especially for men, no shorts or jeans, no sports trainers, tied-back long hair and a clean white stock coat, your number neatly pinned on, and a clean stick and board. The board should be plain white with no personal advertising, although traditionally manufacturers' logos are permissible. Home-made boards are allowed, but make sure they are not too cumbersome or heavy or you will not be able to use them to good effect.

Begin preparing your pig with enough time to enter the ring the minute your class number is called. When you enter the ring, keep on the pig's left so you're not blocking the judge's view and keep your pig moving forwards in a clockwise direction, keeping your board in your left hand and stick in your right. When approached by the judge, stop your pig as best you can, using your board, for them to have a close examination, and answer all their questions such as age, breeding, etc., but don't ask any yourself. Once your pig has been examined, then you can exploit empty corners to 'rest' your pig by using the board on the open side to keep your pig still, but only when you are confident the judge is not looking at you.

After your pig has been thoroughly inspected, don't think they will not look at them again: they will, especially if the judge is deciding between two pigs for a place. In a large class the judge may ask you to leave the ring after the initial judging so they can concentrate on the assessment of other pigs. Don't think you have lost and go off for a burger until the class is over, as you may be asked back in. The judge or steward will give out the rosettes and prize cards to the placed pigs. If you win your class, you will automatically be put forward for the best of breed and the breed/interbreed championships, and if you have been sprinkled with fairy dust, the supreme pig championship (Figure 14.8).

Take your time practising your handling skills. It *will* go wrong sometimes, and even 'professionals' can be outwitted by a clever porcine. Pig people are a truly friendly bunch, so whether you need equipment, advice or an extra pair of hands, I guarantee someone will volunteer to help.

FIGURE 14.8 Winning a supreme champion title makes all the effort worth while.

EPILOGUE

The Commuter Pig Keeper was never intended to be a padded out-veterinary manual but a comprehensive guide to small-scale pig keeping in the UK and the USA, via a collation of my personal experience, anecdotal evidence by fellow pig breeders and keepers and published research.

I also wanted to provide key information in greater depth to assist not only newcomers to pig keeping but also those wishing to expand upon their knowledge and/or venture into the pedigree world or into breeding, for example. I purposefully kept the veterinary information to an overview and aid to steer you in the right direction when seeking further advice. In my opinion this book should be co-owned with a broad pig-specific veterinary manual written by experts. I have recommended three in the Bibliography.

I would like to think that newcomers to pig keeping in both the UK and USA could now go away and start keeping pigs without breaking any laws by accident and keep their pigs in either good health or at least know what to do when things go wrong.

Those with time pressures such as jobs can now see how they could fit the keeping of pigs into their lives and, if desired, give themselves a sizable slice of the 'good life'.

For those who have already embarked on a pig-keeping journey of some sort, I hope that I have given greater depth of information than is currently available in one book, with easily understandable explanations in key areas.

I have thoroughly enjoyed researching and writing *The Commuter Pig Keeper* and hope it becomes a useful reference book for newcomers and experienced pig keepers alike.

BIBLIOGRAPHY AND USEFUL CONTACTS

SPECIALIST VETERINARY REFERENCE BOOKS

If you keep more than a couple of fattening pigs per year or breed pigs, then there are three veterinary text books that you ought to consider having on the shelf. All three have become my bibles.

Managing Pig Health: A Reference for the Farm (2nd Edition) written by Michael Muirhead, Thomas Alexander and edited by John Carr, is a complete veterinary textbook that goes into significant detail and description. 5M Publishing, ISBN 978-0-9555011-5-9.

Pig Ailments: Recognition and Treatment, written by Mark White, is a clearly written, non-technical veterinary book with lots of supportive photographs to aid diagnosis. The Crowood Press, ISBN 1-86126-787-8.

Pig Veterinary Society: The Casualty Pig, interim revision 2013, ISSN 0956-0939. This booklet details what to do and when in various potential casualty scenarios. It is available to buy as a hard copy booklet or free of charge to download online in pdf format from www.pigvetsoc.org.uk.

FURTHER GENERAL READING

The Haynes Pig Manual: Complete Step-by-step Guide to Keeping Pigs by Liz Shankland. This is a highly illustrated book showing you photographically each step in any described process. Haynes, ISBN 978-0857330260.

Pigs: A Guide to Management (2nd Edition) by Neville Beyron. A highly informative book that although written from more of a commercial angle than *The Commuter Pig Keeper* offers insight that is applicable to all pig breeders and keepers, especially those looking to run a commercial herd. The Crowood Press, ISBN 978-1-84797-752-6.

Pig Production: What the Text Books Don't Tell You by John Gadd. A well-known pig expert and author in both the UK and USA gives advice with an emphasis on econometrics (cost effectiveness) in pig production. Nottingham University Press, ISBN 1-904761-21-6.

On Farm Monitoring of Pig Welfare. Written by multiple authors and edited by A. Velarde and R. Geers. A review of published pig welfare research concentrated into one book. Wageningen Academic Publishers, ISBN 978-90-8686-025-8.

USEFUL CONTACTS

Regulations UK

Defra website	www.gov.uk/government/organisations/department-for-environment-food-rural-affairs
Rural Payments Agency (RPA) England Wales Scotland	www.gov.uk/government/organisations/rural-payments-agency //gov.wales/topics/environmentcountryside/helpandadvice/county-parish-holding-numbers/?lang=en www.scotland.gov.uk/Publications/2010/02/08120157/14 RPA helpline 03000 200 301
Animal and Plant Health Agency (APHA)	www.gov.uk/government/organisations/animal-and-plant-health-agency APHA helpline 03000 200 301
Animal Movement Licensing	www.eaml2.org.uk www.scoteid.com/node/4854
Welfare of Animals in Transit (WATO)	ww.gov.uk/government/uploads/system/uploads/attachment_data/file/193680/pb13550-wato-guidance.pdf
Welfare of Animals in Transit (WAIT) examination bodies	LANTRA: www.lantra-awards.co.uk City and Guilds: www.nptc.org.uk Scottish Skills Testing Service: www.sayfc.org/ssts

Regulations USA

Federal animals and animal product legislation including swine rules (in full) 2002	www.gpo.gov/fdsys/pkg/CFR-2002-title9-vol1/pdf/CFR-2002-title9-vol1.pdf
Final swine garbage feeding rule 2009	www.aphis.usda.gov/animal_health/animal_dis_spec/swine/downloads/shp_garbage_feeding_final_rule.pdf
Federal swine disease information	www.aphis.usda.gov/wps/portal/aphis/ourfocus/animalhealth/sa_animal_disease_information/sa_swine_health
Official AIN/840 ear tags explained	www.aphis.usda.gov/traceability/downloads/AIN_device_list.pdf

Approved Ear Tag/Identification

Fearing International	www.fearing.co.uk
Dalton Ltd	www.dalton.co.uk
Ketchum Manufacturing	www.ketchums.co.uk
Id & Trace	www.idandtrace.com
Allflex Europe Ltd	www.allflex.co.uk
Allflex USA, Inc.	www.allflexusa.com
Destron Fearing (USA)	www.destronfearing.com
Y-Tex Corporation (USA)	www.Ytex.com

Breed Clubs and Associations UK

British Pig Association	www.britishpigs.org
Rare Breed Survival Trust	www.rbst.org.uk
British Kunekune Pig Society	www.britishkunekunesociety.org.uk
British Lop Pig Society	www.britishloppig.org.uk
British Saddleback Breeders Club	www.saddlebacks.org.uk
Gloucestershire Old Spots Pig Breeders Club	www.gospbc.co.uk
Tamworth Breeders Club	www.tamworthbreedersclub.co.uk
Berkshire Pig Breeders Club	www.berkshirepigs.org.uk
Large Black Pig Breeders Club	www.largeblackpigs.org.uk
Oxford, Sandy and Black Pig Society	www.oxfordsandypigs.co.uk
Middle White Pig Breeders Club	www.middlewhite.co.uk
Pedigree Welsh Pig Society	www.pedigreewelsh.com
Mangalitza	via www.britishpigs.org
Duroc	via www.britishpigs.org
Hampshire	via www.britishpigs.org
British Landrace	via www.britishpigs.org
Large White	via www.britishpigs.org
Piétrain	via www.britishpigs.org

Breed Registries USA

National Swine Registry (NSR)	www.nationalswine.com
Certified Pedigreed Swine (CPS)	www.cpsswine.com
American Berkshire Association (ABA)	www.americanberkshire.com
The Livestock Conservancy (TLC)	www.livestockconservancy.org
Gloucestershire Old Spots of America (GOSA)	www.gosamerica.org
Gloucestershire Old Spots Pig Breeders United (GOSPBU)	www.gospbu.org
American Guinea Hog Association (AGHA)	www.guineahogs.org
National Hereford Hog Association (NHHA)	www.nationalherefordhogassociation.com
Juliana Pig Association and Registry	www.julianapig.com
American Kunekune Pig Society (AKKPS)	www.americankunekunepigsociety.com
American Kunekune Pig Registry (AKKPR)	www.americankunekunepigregistry.com
Large Black Hog Association (LBHA)	www.largeblackhogassociation.org
North American Large Black Pig Registry	No website
American Mangalitsa Breeders Association	www.americanmangalitsa.com
American Mulefoot Breeders Association	www.americanmulefoot.com
American Mulefoot Hog Association and Registry	www.mulefootpigs.tripod.com
Red Wattle Hog Association	www.redwattle.com
Tamworth Swine Association	www.tamworthswine.org

Organic Pig Production

UK	www.soilassociation.org/whatisorganic/organicanimals/pigs www.soilassociation.org/farmersgrowers/technicalinformation/pigs
USA	https://attra.ncat.org/publication.html#organic (Document IP185)

Commercial Organisations with Useful Information, from Training Opportunities and Current Disease Situations Through to Pork Recipes

ORGANISATION	WEBSITE
Agricultural and Horticultural Development Board – Pork (UK)	http://pork.ahdb.org.uk/
The Pig Site (UK, USA)	www.thepigsite.com
Love Pork: Pork Facts, Pork Cuts, Recipes and Healthy eating (UK)	www.lovepork.co.uk
National Farmers Union (UK)	www.nfuonline.com
Pig Veterinary Society (UK)	www.pigvetsoc.org.uk
Pork Checkoff (USA)	www.pork.org
National Pork Producers Council (USA)	www.nppc.org

Artificial Insemination

COMPANY	WEBSITE
Traditional breeds (UK)	www.deerpark-pigs.com www.britishpigs.org.uk/AI%20instructions%202007.pdf
Modern breeds (UK)	www.deerpark-pigs.com www.acmc.co.uk www.jsrgenetics.com www.rattlerow.co.uk
List of modern and heritage breed suppliers (USA) Also contact the specific breed registries	www.thepigsite.com/directory/2/north-america/2/semen-suppliers-artificial-insemination www.heritageswine-ai.com www.showpigs.com/index.php http://internationalboarsemen.us www.shipleyswine.com www.swinegenetics.com

Abattoir and Regulations

ACTIVITY TYPE	WEB LINKS
Approved abattoirs and private on-farm slaughter regulations (UK)	www.food.gov.uk Links to Wales, Scotland and Northern Ireland
Approved abattoir/plant list (USA)	www.fsis.usda.gov/wps/wcm/connect/ a5c2b5c8-92e0-4565-8999-f2fb75bfdb05/ MPI_Directory_Establishment_Number. pdf?MOD=AJPERES

Red meat mobile abattoirs (USA)	www.extension.org/pages/19781/mobile-slaughterprocessing-units-currently-in-operation#.VbzgvJXbLIU
Private on-farm slaughter regulations (USA)	www.aasv.org/aasv/euthanasia.pdf www.ers.usda.gov/media/820188/ldpm216-01.pdf

Shows and Exhibitions

SHOW LINKS	WEBSITE
List of all BPA affiliated and accredited shows in the UK	www.britishpigs.org
All agricultural shows in the UK	www.asao.co.uk
National Swine Registry shows (USA)	http://nationalswine.com/shows http://nationalswine.com/shows/open_shows/extravaganza/extravaganza_rules.php

INDEX